Springer Theses

Recognizing Outstanding Ph.D. Research

Aims and Scope

The series "Springer Theses" brings together a selection of the very best Ph.D. theses from around the world and across the physical sciences. Nominated and endorsed by two recognized specialists, each published volume has been selected for its scientific excellence and the high impact of its contents for the pertinent field of research. For greater accessibility to non-specialists, the published versions include an extended introduction, as well as a foreword by the student's supervisor explaining the special relevance of the work for the field. As a whole, the series will provide a valuable resource both for newcomers to the research fields described, and for other scientists seeking detailed background information on special questions. Finally, it provides an accredited documentation of the valuable contributions made by today's younger generation of scientists.

Theses are accepted into the series by invited nomination only and must fulfill all of the following criteria

- They must be written in good English.
- The topic should fall within the confines of Chemistry, Physics, Earth Sciences, Engineering and related interdisciplinary fields such as Materials, Nanoscience, Chemical Engineering, Complex Systems and Biophysics.
- The work reported in the thesis must represent a significant scientific advance.
- If the thesis includes previously published material, permission to reproduce this must be gained from the respective copyright holder.
- They must have been examined and passed during the 12 months prior to nomination.
- Each thesis should include a foreword by the supervisor outlining the significance of its content.
- The theses should have a clearly defined structure including an introduction accessible to scientists not expert in that particular field.

More information about this series at http://www.springer.com/series/8790

Giuseppe Di Domenico

Electro-optic Photonic Circuits

From Linear and Nonlinear Waves in Nanodisordered Photorefractive Ferroelectrics

Doctoral Thesis accepted by
the University of Rome La Sapienza, Rome, Italy

Springer

Author
Dr. Giuseppe Di Domenico
School of Electrical Engineering
Tel Aviv University
Tel Aviv, Israel

Supervisor
Prof. Eugenio Del Re
Physics Department
University of Rome La Sapienza
Rome, Italy

ISSN 2190-5053 ISSN 2190-5061 (electronic)
Springer Theses
ISBN 978-3-030-23188-0 ISBN 978-3-030-23189-7 (eBook)
https://doi.org/10.1007/978-3-030-23189-7

This Springer imprint is published by the registered company Springer Nature Switzerland AG
The registered company address is: Gewerbestrasse 11, 6330 Cham, Switzerland

This work is for Giulia,
who deserves a Ph.D. in being the best.

Supervisor's Foreword

An electrical circuit can be seen as a family of electric devices that act on flowing current; each device is based on different physical principles and realized in different materials; and all are integrated together on a single board. An optical circuit is exactly that a series of single optical devices pooled together to perform a given operation on propagating light. Giuseppe Di Domenico, in this present treatise, describes his exploration and development of a method to realize generic optical operations integrated and miniaturized inside one single bulk material. The starting point of his journey is a rather intriguing setting, a nanodisordered ferroelectric. While his final achievements, such as the ability to integrate an electro-optic Gaussian-to-Bessel beam converted into a miniscule slot of the, so to say, crystalline motherboard, are inspiring to an applicative eye, it is the journey itself that captures our attention. Choosing a material in an optical research initiative sometimes sets you along a wholly unexpected and adventurous path. This is certainly the case for potassium-lithium-tantalate-niobate and potassium-sodium-tantalate-niobate described here. The choice is generally between a well-known material with established industrial pedigree or a newly developed material with still unknown properties, but with the potential to make things work. To achieve a complete optical operation, you require an optical source, detection, guiding, fast modulation and steering, electronic control, polarization control, wavelength selectivity, and nonlinearity, all miniaturized in a full three-dimensional volume, where light has its edge on electronic circuits. Aside from light generation and detection, all the other ingredients are found in nanodisordered ferroelectrics, that is, ferroelectrics achieved by mixing different crystals into a single perovskite structure, a crystal that is actually a solid solution. A particularly enticing material has been developed by Aharon J. Agranat and his group at the Hebrew University of Jerusalem. This is a material that has a room-temperature phase transition and maintains its optical quality, so that it can be used in proximity of its dielectric anomaly at the Curie point. Here, the material manifests electro-holography, electro-optic waveguides, spatial solitons, scale-free optics, giant refraction, huge second-harmonic generation, and a hereto unknown ferroelectric phase, a ferroelectric supercrystal composed of hypervortices. While all these add up to a multifunctional material, a candidate to host generic optical operations, they also lead to a rich playground for the observation and discovery of new phenomena, such as the first observation of the supercrystal phase and the intriguing detection of a nonlinearity able to generate an intrinsic negative mass, that is, a Schroedinger wave equation with a negative mass term not in the delocalized Bloch modes, as occurs for holes in semiconductors, but in the actual localized optical wave.

Rome, Italy
June 2019

Prof. Eugenio Del Re

Abstract

The work presented in this thesis addresses different aspects of three main physical issues belonging to the field of nonlinear optics, quantum optics, and optical microscopy. We analyze how photorefraction can be used to photoinduce a tapered fiber index of refraction patterns in the bulk of nanodisordered crystals, and we observe how these patterns are able to modulate the phase of Gaussian beams converting them to Bessel–Gauss beams, enhancing their depth of field and their ability to self-heal after an obstacle. These properties suggest the use of Bessel beam in microscopy. In our investigations, we proposed and experimentally demonstrated, in turbid media, the idea of using the interference between multiple Bessel beams to generate a light field that is non-diffracting and self-healing, but also localized along the propagation axis. Our study on superimposed Bessel beams reveals how the interference between their side lobes has the overall effect of reducing the amount of energy possessed by the beam outer structures, practically enhancing their localization in the radial direction as well as in the axial. At present, we are studying how to implement these findings in a light sheet microscope to improve optical sectioning. Also described in this thesis are a number of intriguing experiments carried out on disordered ferroelectrics and their giant response, including negative intrinsic mass dynamics, ferroelectric supercrystals, rogue wave dynamics driven by enhanced disorder, and the first evidence of spatial optical turbulence. Lastly, relying on the necessarily reversible nature of the microscopic process, we demonstrate how a single photon is not able to entangle two distant atoms because of conservation laws, clarifying the long-standing debate on the nature of single-photon non-locality and introducing fundamental limitation, in the use of linear optics for quantum technology.

The thesis is organized as follows: In Chap. 1 we present the basic mechanism of photorefractive spatial solitons and the phenomena involved in soliton formation, whereas in Chap. 2 we describe the basic concept of optical microscopy. Chapter 3 is dedicated to the experimental observation of a miniaturized device that can convert a Gaussian beam into a Bessel beam with low response time and using volume integrable techniques. Chapter 4 reports the experimental investigation of a new non-diffracting light-field-generated superposing multiple Bessel Beams, what we term “light droplet.” Further experiments demonstrate that the droplets are self-healing in turbid media, a feature that can potentially allow in vivo imaging of thick specimens in a backscattering microscope configuration. In Chap. 5, we continue the experiments on “light droplets” and Bessel beam superposition by showing how the interference of their side lobe leads to a reduction of the off-axial intensity. In Chap. 6, we discuss theoretically the idea of entangling two distant systems by means of a single particle. Our findings show how, from first principles, quantum mechanics prevents the possibility of entangling the two systems in any useful way. In Chap. 7, the discovery of a spontaneous polarization supercrystals in microstructured samples of potassium-lithium-tantalate-niobate (KLTN) is described. This polarization domain structure is three-dimensional, has a micrometric period, and affects the light propagating through it by spatially separating its polarization components. In Chap. 8, we describe experiments, theory, and numerics that show how anti-diffracting nonlinear waves evolving into an optical potential made by an integrated slab waveguide give rise to the dynamics of a negative mass quantum particle. Chapter 9 reports direct evidence of turbulent transitions in optical wave propagation. The transition occurs as the disordered hosting material passes from being linear to being extremely nonlinear, a regime that unveils the emergence of concomitant rogue waves. The control of these extreme events through spatial incoherence is also experimentally demonstrated.

Preface

The research contained in this thesis was carried out under the supervision of Eugenio Del Re, in collaboration with Fabrizio Di Mei, Davide Pierangeli, Giuseppe Antonacci, Paolo Di Porto, Salvatore Silvestri, Jacopo Parravicini, and Aharon J. Agranat. The experimental activity has been implemented in the laboratory of the Physics Department of the University of Rome "La Sapienza" (S005-S008, Fermi building) thanks to funding from the FIRB Futuro in Ricerca project grants PHOCOSRBFR08E7VA, the PRIN project 2012BFNWZ2, and Sapienza 2014–2015 Awards projects. The support from the Italian Institute of Technology IIT@Sapienza (Center for Life Nano Science) is also acknowledged.

Tel Aviv, Israel | Dr. Giuseppe Di Domenico

Publications Related to This Thesis

Papers published by the candidate during the three-year Ph.D. course:

Davide Pierangeli, Mario Ferraro, Fabrizio Di Mei, Giuseppe Di Domenico, C.E.M. de Oliveira, Aharon J. Agranat, Eugenio Del Re. *Super-crystals in composite ferroelectrics*. Nat. Comm. **7**, 10674 (2016).

Fabrizio Di Mei, Piergiorgio Caramazza, Davide Pierangeli, Giuseppe Di Domenico, H. Ilan, Aharon J. Agranat, Paolo Di Porto, Eugenio DelRe. *Intrinsic negative mass from nonlinearity*. Phys. Rev. Lett. **116**, 153902 (2016).

Davide Pierangeli, Fabrizio Di Mei, Giuseppe Di Domenico, Aharon J. Agranat, Claudio Conti, Eugenio DelRe. *Turbulent transitions in optical wave propagation*. Phys. Rev. Lett. **117**, 183902 (2016).

Davide Pierangeli, Gabriella Musarra, Fabrizio Di Mei, Giuseppe Di Domenico, Aharon J. Agranat, Claudio Conti, Eugenio DelRe. *Enhancing optical extreme events through input wave disorder*. Phys. Rev. A **94**, 063833 (2016).

Giuseppe Antonacci, Simone De Panfilis Giuseppe Di Domenico, Eugenio DelRe, Giancarlo Ruocco. *Breaking the Contrast Limit in Single-Pass Fabry-Pérot Spectrometers*. Phys. Rev. App. **6**, 054020 (2016).

Giuseppe Antonacci, Giuseppe Di Domenico, Salvatore Silvestri, Eugenio DelRe, Giancarlo Ruocco. *Diffraction-free light droplets for axially-resolved volume imaging*. Sci. Rep. **7**, 17 (2017).

Giuseppe Di Domenico, Jacopo Parravicini, Giuseppe Antonacci, Salvatore Silvestri, Aharon J. Agranat, Eugenio DelRe. *Miniaturized photogenerated electro-optic axicon lens Gaussian-to-Bessel beam conversion*. Appl. Opt. **56**, 2908 (2017).

Mario Ferraro, Davide Pierangeli, Mariano Flammini, Giuseppe Di Domenico, Ludovica Falsi, Fabrizio Di Mei, Aharon J. Agranat, Eugenio DelRe. *Observation of polarization-maintaining light propagation in depoled compositionally disordered ferroelectrics*. Opt. Lett. **42**, 3856 (2017).

Giuseppe Di Domenico, Giancarlo Ruocco, Cristina Colosi, Eugenio DelRe, Giuseppe Antonacci. *Self-suppression of Bessel beam side lobes for high-contrast light sheet microscopy*. Sci. Rep. **8**, 1 (2018).

Mariano Flammini, Giuseppe Di Domenico, Davide Pierangeli, Fabrizio Di Mei, Aharon J. Agranat and Eugenio DelRe. *Observation of Bessel-beam self-trapping*. Phys. Rev. A. **98**, 3 (2018).

Papers presented by the candidate at international conferences:

Davide Pierangeli, Mario Ferraro, Fabrizio Di Mei, Giuseppe Di Domenico, C.E.M. de Oliveira, Aharon J. Agranat, Eugenio DelRe. *Spontaneous photonic super-crystals in composite ferroelectrics*. Conference on Lasers and Electro-Optics (CLEO), San Francisco CA, USA (2016).

Papers prepared by the candidate and currently under review process in journals:

Paolo Di Porto, Giuseppe Di Domenico, Fabrizio Di Mei, Bruno Crosignani, Eugenio DelRe. *Microscopic reversibility and no entanglement produced by a single particle on distant systems*. Submitted to Physical Review Letters.

Acknowledgements

Many people have helped me in one way or another to write this thesis, and unavoidably there will be a number among them, whose valuable contributions will go unattributed, owing to congenital disorganization and forgetfulness on my part. Let me first express my special thanks (and also apologies) to these people. Apart from them for more specific help and assistance, I can directly express my gratitude.

I would like to express my sincere gratitude to my advisor Prof. Eugenio Del Re for the continuous support he gave me during my Ph.D. study and related research, and for his patience, motivation, and knowledge. His guidance helped me during the researching and writing of my thesis. In addition to our academic collaboration, I greatly value the personal rapport that Eugenio and I have forged over the years. I quite simply cannot imagine a better advisor for my studies.

Besides my advisor, my sincere thanks also go to my fellow labmates Fabrizio Di Mei, Davide Pierangeli, Mariano Flammini, Ludovica Falsi, and Giuseppe Antonacci for their time, for their valuable feedback, for the stimulating discussions, and lastly for all the fun we have had over the last few years.

Finally, my gratitude goes to my family: my parents Luigi and Antonietta and my sisters Elena end Giulia, not merely for their support, but for their continued love and care, and their deep understanding and sensitivity, despite the seemingly endless years of having a son/brother who was mentally only half present.

Of course, no acknowledgments would be complete without acknowledging with tremendous and deep thanks my friends who have helped me make the common life less common.

Thank you.

April 2019

Dr. Giuseppe Di Domenico
La Sapienza University

Contents

Chapter 1
Introduction to Nonlinear Optics in Photorefractive Media

In this chapter nonlinear optical beams are introduced and specialized to spatial solitons in photorefractive media. In particular, we present the electro-optic effect, ferroelectricity, relaxor ferroelectrics, the mechanism of photorefraction, and the physics underlying photorefractive solitons.

1.1 The Electro-optic Effect

The electro-optic effect refers to a phenomenon in which an external (static or low-frequency) electric field changes the optical properties of a medium. Several applications can take advantage of this phenomenon, primarily information technology where electrical signals can be transformed into either a phase or an intensity change in a light signal with high modulation speed [1–3]. Other examples are laser Q-switches [4], laser pulse shaping [5], modulating retroreflectors [6], tunable optical microresonators [7, 8], liquid crystal technology [9] and Pockels cells. Finally electro-optic detection systems have been successfully applied to study the time structure of ultrashort electron bunches [10, 11]. In an electro-optic crystal, the refractive index becomes a function of the external field. At an atomic level, an electric field applied to certain crystals causes a redistribution of bond charges and slight deformation of the crystal lattice [12]. In general, these alterations are not isotropic, that is, the changes vary with direction in the crystal. Therefore, the impermeability tensor Eq. 1.1 changes accordingly. Crystals lacking a center of symmetry are non-centrosymmetric and exhibit a linear (Pockels) electro-optic effect, the changes in the impermeability tensor elements are linear in the applied electric field. On the other hand, all crystals exhibit a quadratic (Kerr) electro-optic effect where the changes in the impermeability tensor elements are quadratic in the applied field. When the linear effect is present, it generally dominates over the quadratic effect. In general, the impermeability tensor can be expressed through a power series [13]

G. Di Domenico, *Electro-optic Photonic Circuits*, Springer Theses,
https://doi.org/10.1007/978-3-030-23189-7_1

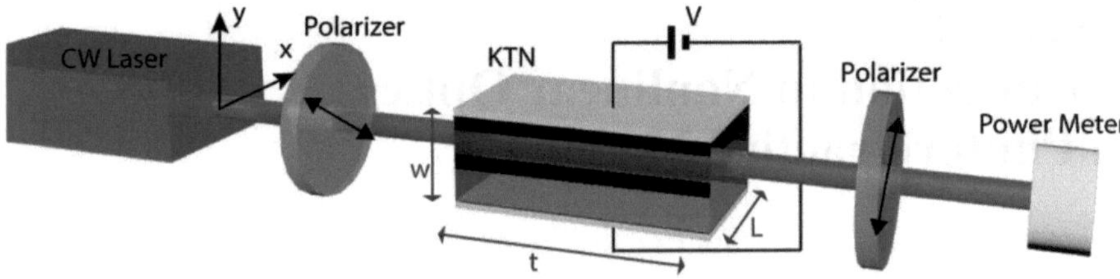

Fig. 1.1 Illustration of a cross polarizer set-up

$$\Delta\left(\frac{1}{n^2}\right)_{ij} = \sum_k f_{ijk}P_k + \sum_{kl} g_{ijkl}P_kP_l = \sum_k r_{ijk}E_k + \sum_{kl} s_{ijkl}E_kE_l, \tag{1.1}$$

where E is the applied electric field, P is the polarization field vector, the constants r_{ijk} and f_{ijk} are the Pokels coefficients and s_{ijkl} and g_{ijkl} are the Kerr ones defined as

$$f_{ijk} = \frac{1}{2}\left(\frac{\partial\left(\frac{1}{\Delta n_{ij}}\right)}{\partial E_k}\right)_{E=0} = \frac{r_{ijk}}{\varepsilon_k - \varepsilon_0},$$
$$s_{ijkl} = \frac{1}{2}\left(\frac{\partial^2\left(\frac{1}{\Delta n_{ij}}\right)}{\partial E_k \partial E_l}\right)_{E=0} = \frac{g_{ijkl}}{(\varepsilon_k - \varepsilon_0)(\varepsilon_l - \varepsilon_0)}. \tag{1.2}$$

The electro-optic response of crystals can be investigated with the setup in Fig. 1.1 [14], this is the simplest detection setup, usually referred to as the crossed polarizers. It is alined so that the two polarizer are oriented in $\hat{x} + \hat{y}$ and $\hat{x} - \hat{y}$ direction (assuming light incidence perpendicular to xy-plane). From the Malus law the output intensity I of a plane wave input I_0 propagating along z is given by

$$I = I_0 sin^2\left(\frac{\Delta\phi}{2} + \phi_0\right), \tag{1.3}$$

where $\Delta\phi$ is the relative phase shift between the optical component polarized along $\hat{x}$ and $\hat{y}$ and ϕ_0 is an eventual residual birefringence, which can also be compensated by a quarter-wave plate between the crystal and the final polarizer. This phase shift is accumulated during the propagation inside the crystal due to the electro-optically induced birefringence and is connected to the Δn through the relation

$$\Delta\phi = \Delta n(2\pi/\lambda)L, \tag{1.4}$$

with L the length of the crystal along the propagation direction. In most cases the electro-optic effect is observed in crystals having a dielectric constant ϵ strongly dependent on the temperature. For this reason, in Eq. 1.1, it is more appropriate to consider the expansion in terms of polarization. For quadratic electro-optic effect we have

$$\Delta\left(\frac{1}{n^2}\right)_{ij} = \sum_{kl} g_{ijkl} P_k P_l, \qquad (1.5)$$

where g_{ijkl} are the elements of the quadratic electro-optical tensor expressed through the $\boldsymbol{P}$ components which, in distinction to s_{ijkl}, are independent from temperature. The response assumes a scalar form when the optical axes are chosen as reference system, so that g_{ijkl} is diagonal and the index of refraction variation $\Delta n(E)$ becomes

$$\Delta n(E) = -\frac{1}{2} n^3 g_{eff} \epsilon_0^2 (\epsilon_r - 1)^2 E^2 \qquad (1.6)$$

This relation can be applied to the setup Fig. 1.1 (where the electric field is applied along the direction $\hat{y}$) considering $g_{eff} = g_{11} - g_{12}$. With the cross polarizer set-up is possible to trace back the value of P as a function of external electric field from the measure of the output intensity through the Eqs. 1.3, 1.4, 1.6.

1.2 Relaxors

Some disordered ferroelectric crystals, especially perovskites, exhibit complex macroscopic response, with many analogies with dipolar glass physics. These are the so called "Relaxor" ferro-electrics and are used in the design of high performance ultrasonic and sonar transducers [15–17], as energy harvesters [18], and in aerospace industries [19]. In a Relaxor crystal the electro-optic response can be greatly enhanced, becoming a giant effect near the para to ferro-electric transition. In addition, the introduction of compositional disorder on the nanoscale leads to a dispersive dielectric susceptibility that manifests thermal, electric field, and strain hysteresis along with anomalous relaxation times [20, 21]. The presence of different compounds introduces, for specific composition concentrations, competing structural phases that, at the morphotropic phase boundaries [22, 23], such as low-symmetry bridging phases and ferroelectric-antiferroelectric ordering boundaries [24], leading to unique polarization properties, examples being anomalously large capacitance, piezoelectric coefficients an order of magnitude larger than traditional ferroelectric ceramics [25, 26] and have anomalously material loss [27, 28].

The physical mechanism behind Relaxor behavior has been the subject of extensive research over the last decades, and is at present not fully understood. Several models have been proposed over the years to explain relaxor behavior [29]. In many respects, the unique properties of the disordered ferroelectric state can be modelled as arising from a network of randomly interacting nanoscale polar domains (polar-nanoregions, PNRs) embedded in a highly polarizable medium [30–34]. The ferroelectric relaxors have two or more cations occupying equivalent crystallographic sites in the lattice structure. This creates the PNRs, the size of which varies with temperature, induced by the dopant ions. Random fields theory [35–40] considers ferroelectric relaxor as an intermediate state between dipole glasses and normal ferroelectrics. In contrast to dipolar glasses, where elementary dipolar moments exist on the atomic scale, the relaxor state is characterized by the presence of nanoscale polar clusters of variable sizes. Nanoscale polar clusters are spherical and interact randomly. Although the microscopic origin of these PNRs and the role played by random fields is still an open question, it is established that they can lead to dipolar-glasses with non-ergodic properties when appropriately supercooled [41]. In fact, dipolar dynamics in some relaxor ferroelectric crystals is characterized by the so-called freezing temperature, a temperature at which the dielectric relaxation time diverges and polarization fluctuations are quenched [42–45]. Percolation of PNRs has been proposed as the physical mechanism underlying the dipolar-glass state [46, 47]. Moreover, in proximity of the Curie temperature, in the nominally paraelectric (cubic) phase, PNRs greatly affect optical birifringence [48, 49] and electro-optical response leading to giant electro-optical coefficients [50, 51] and depolarization effects [52].

1.3 Photorefractive Effect

The photorefractive effect is a process by which light generates a space-charge field that locally modifies the index of refraction ([10, 53] for the original observation). The mechanism can be summarized as follows

1. Light propagating through the medium can be absorbed by impurities, exciting the transitions of electrons to the conduction band.
2. Free electrons in the conduction band can now diffuse under the effect of the local electron density gradient (diffusion) or be driven by an external electric field (drift)
3. Spatially redistributed electrons recombine with holes at positions different from their photo-excitation sites.

In an electro-optic medium Sect. 1.1, the space charge field produces a change in the refractive index able to affect the optical beam which activated the whole process. The result is a nonlinear optical effect [54].

An important feature of the photorefractive effect effect is that it is cumulative, that is, the response can build up in time as more and more charges are displaced from the illuminated regions to the darker one. The result is that even extremely low optical intensities (of the order of 1 W/cm^2) can gives rise to a strong optical nonlinearity at the expense of a proportionally slower response. Photorefraction can support a large class of nonlinear wave dynamics [53] and different kinds of spatial optical solitons at intensities that are much smaller than those required to observe other nonlinear phenomena (like the well-known Kerr Effect whose intensity threshold is of the order of 10^6 W/cm^2).

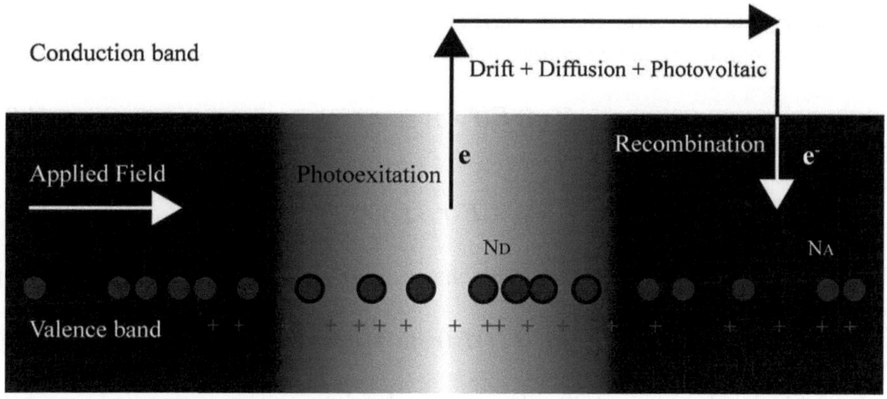

Fig. 1.2 Band scheme of a single level photorefractive process

1.3.1 Band Transport Model

In the simplest possible photorefractive model [55] a single level of trapping sites is introduced between the valence and the conduction band as shown in Fig. 1.2. Consider a dielectric medium with deep donor impurities density N_D and acceptor impurities density N_A, with $N_D \gg N_A$. Light of wavelength λ excites locally free carriers that are subject to drift and diffusion fields and then recombines with acceptor impurities. The result is a difference in charge distribution between bright and dark regions in the dielectric, with a dynamic described by the rate equation

$$\frac{\partial N_D^+(\boldsymbol{r},t)}{\partial t} = (\beta + sI(\boldsymbol{r},t))[N_D - N_D^+(\boldsymbol{r},t)] - \gamma N_D^+(\boldsymbol{r},t)N_e(\boldsymbol{r},t), \tag{1.7}$$

where β is the thermal excitation constant, s the photoexcitation crosssection, γ the donor-electron recombination factor, I (r, t) the intensity of light and $N_e(r,t)$ the number of electrons excited in conduction band, $N_D(r,t)$ and $N_D^+(r,t)$ are the spatial donor impurity densities, respectively total and ionized. The total space charge density is given by

$$\rho(\boldsymbol{r},t) = q\left[N_D^+(\boldsymbol{r},t) - N_A - N_e(\boldsymbol{r},t)\right], \tag{1.8}$$

where q is the electron charge and N_A is the spatial acceptor impurity densities. In standard photorefractives the condition $N_D \gg N_A$ is fulfilled ($N_D \approx 10^{18} \div 10^{19}$ cm^{-3} and $N_A \approx 10^{16} \div 10^{17}$ cm^{-3}) (at room temperature and in the absence of light) a fraction of donor impurities is ionized since donors provide electrons for the lowest-energy levels associated with acceptors which, consequently, are wholly ionized. The number of electrons in the conduction band is negligible because

$K_B T \ll \epsilon_d$ where ϵ_d is the energy-gap between the donor levels and the bottom of the conduction band. The space-charge field in the crystal satisfies Maxwell equations

$$\nabla \cdot [\epsilon \boldsymbol{E}(\boldsymbol{r}, t)] = \rho(\boldsymbol{r}, t), \tag{1.9}$$

and

$$\nabla \times \boldsymbol{E}(\boldsymbol{r}, t) = 0, \tag{1.10}$$

where ϵ is the dielectric constant. The charge migration is described by the current density

$$\boldsymbol{J}(\boldsymbol{r}, t) = q\mu N_e(\boldsymbol{r}, t)\boldsymbol{E}(\boldsymbol{r}, t) + \mu K_B T \nabla N_e(\boldsymbol{r}, t), \tag{1.11}$$

where μ is the electron mobility. The current density takes into account charge drift in an applied electric field and charge diffusion acting in a non uniform spatial charge density. For our purposes the photo-voltaic term is negligible. Charge and current density are coupled by the continuity equation

$$\nabla \cdot \boldsymbol{J}(\boldsymbol{r}, t) = -\frac{\partial \rho(\boldsymbol{r}, t)}{\partial t}, \tag{1.12}$$

The solution for the set of equations from Eqs. 1.7 to 1.12 gives the field $E(r, t)$ that rules the nonlinearity acting on the optical beam, the resulting system however is not analytically solvable [56–58].

1.3.2 *Space Charge Field*

To derive the space charge field in the general, non stationary, case we first render explicit the value of N_D^+ from Eqs. 1.8 and 1.9

$$N_D^+ = N_e + N_A \left[1 + \nabla \left(\frac{\epsilon \boldsymbol{E}}{q N_A}\right)\right], \tag{1.13}$$

then we observe that in Eq. 1.7, the differential $\frac{\partial N_D^+(\boldsymbol{r},t)}{\partial t}$ reaches its steady state with a time-scale of the order of 100 ns (proportional to $\frac{1}{\beta}$) whereas the dielectric relaxation time $\tau_D = \frac{\epsilon}{q\mu N_e}$, the average time needed to screen an arbitrary electric field distribution inside the material, is typically of the order of seconds. Therefore, it is a good approximation to consider $\frac{\partial N_D^+(\boldsymbol{r},t)}{\partial t} \approx 0$ even in the non-stationary case. The density of electrons excited in the conduction band can be evaluated from Eqs. 1.7 and 1.13 obtaining

$$N_e^2 + N_e\left[1+\frac{s}{\beta}I + N_A\left[1+\nabla\left(\frac{\epsilon \boldsymbol{E}}{qN_A}\right)\right] - N_A\left(\frac{\beta+sI}{\gamma}\right)\left[\frac{N_D-N_A}{N_A} - \nabla\left(\frac{\epsilon \boldsymbol{E}}{qN_A}\right)\right]\right] = 0, \tag{1.14}$$

the only solution which has physical meaning is the positive one

$$N_e = \frac{\beta}{2\gamma}\left[\left(1+\frac{s}{\beta}I\right) + \frac{\gamma N_A}{\beta}\left(1+\nabla\cdot\left(\frac{\epsilon \boldsymbol{E}}{qN_A}\right)\right)\right]\cdot$$
$$\cdot\left[-1+\sqrt{1+\frac{4N_A\frac{\beta}{\gamma}\left(1+\frac{s}{\beta}I\right)\left[\frac{N_D-N_A}{N_A} - \nabla\cdot\left(\frac{\epsilon \boldsymbol{E}}{qN_A}\right)\right]}{\frac{\beta^2}{\gamma^2}\left[\left(1+\frac{s}{\beta}I\right) + \frac{\gamma N_A}{\beta}\left[1+\nabla\cdot\left(\frac{\epsilon \boldsymbol{E}}{qN_A}\right)\right]\right]^2}}\,\right], \tag{1.15}$$

from Eqs. 1.9, 1.11, 1.12 we easily get:

$$\nabla\cdot\left[\epsilon\frac{\partial \boldsymbol{E}}{\partial t} + q\mu N_e \boldsymbol{E} + \mu K_b T \nabla N_e\right] = 0. \tag{1.16}$$

Here the dielectric constant ϵ is at zero frequency since Eq. 1.16 holds on the time scales where the dielectric response is quasi-static.

These Eqs. 1.15 and 1.16 accurately predict the space-charge electric field $\boldsymbol{E}(\boldsymbol{r}, t)$ produced by a known optical intensity distribution $I(\boldsymbol{r}, t)$ within the photorefractive bulk. Considering now the case $N_e \ll N_A, N_D$, (a condition satisfied in most practical cases), in Eq. 1.13 N_e can be neglected with respect to the second term. From the rate equation, the carriers density in the conduction band reads

$$N_e = \frac{\beta+sI}{\gamma}\left[\frac{N_D - N_D^+}{N_D^+}\right], \tag{1.17}$$

that, through Eq. 1.13 for N_D^+, can be finally expressed as

$$N_e = \frac{\beta+sI}{\gamma}\left[\frac{\frac{N_D-N_A}{N_A} - \nabla\left(\frac{\epsilon \boldsymbol{E}}{qN_A}\right)}{1+\nabla\left(\frac{\epsilon \boldsymbol{E}}{qN_A}\right)}\right]. \tag{1.18}$$

Substituting Eq. 1.18 in Eq. 1.16 and introducing the parameter $\alpha = \frac{N_D-N_A}{N_A}$ ($\alpha \gg 1$), we obtain the implicit equation for the space-charge field as a function of the optical intensity

$$\nabla\cdot\left[\frac{\epsilon\gamma}{q\mu s\alpha}\frac{\partial \boldsymbol{E}}{\partial t} + \boldsymbol{E}(\beta/s+I)\left[\frac{1-\frac{\nabla(\epsilon \boldsymbol{E})}{\alpha q N_A}}{1+\frac{\nabla(\epsilon \boldsymbol{E})}{qN_A}}\right] + \frac{K_b T}{q}\nabla\cdot\left((\beta/s+I)\left[\frac{1-\frac{\nabla(\epsilon \boldsymbol{E})}{\alpha q N_A}}{1+\frac{\nabla(\epsilon \boldsymbol{E})}{qN_A}}\right]\right)\right] = 0, \tag{1.19}$$

the term β/s is called dark intensity I_d and takes into account the thermal contribution to the ionization process. Generally, the thermal contribution is included in the background illumination term I_b. Considering $\alpha \gg 1$ in the quasi-stationary case $\left(\frac{\partial \boldsymbol{E}}{\partial t} \approx 0\right)$ we have

$$\boldsymbol{E}(I_b + I)\frac{1}{1 + \frac{\nabla(\epsilon \boldsymbol{E})}{qN_A}} + \frac{K_b T}{q}\nabla \cdot \left((I_b + I)\frac{1}{1 + \frac{\nabla(\epsilon \boldsymbol{E})}{qN_A}}\right) = g, \tag{1.20}$$

with g a constant value determined by the boundary conditions, such as the external voltage applied to the material in the direction transverse to beam propagation. For instance, $V = 0 \Rightarrow g = 0$, and the space-charge field reduces to

$$\boldsymbol{E} = -\frac{K_b T}{q}\frac{\nabla I}{I_b + I}. \tag{1.21}$$

This is the diffusive electric field used in Chap. 8. In cases where $V \neq 0$, Eq. 1.20 has a non-trivial structure. We will address here the $1 + 1D$ problem (for further studies see [56, 58]). First we normalize the physical quantities introducing

$$Y = \frac{E}{E_0}, \quad Q = \frac{I_b + I}{I_b}, \quad \xi = \frac{x}{x_q} = x\frac{qN_A}{\epsilon E_0}, \tag{1.22}$$

where E_0 is the intensity-independent electric field and x_q is the saturation length. Since the illuminated region $l \ll L$, L being the transverse dimension of the medium, the field E_0 can be approximated as $E_0 \simeq V/L$. Through these variables Eq. 1.20 reads [56]

$$\frac{YQ}{1 + Y'} + a\left[\frac{Q'}{1 + Y'} - \frac{Q}{(1 + Y')^2}Y''\right] = G, \tag{1.23}$$

where we introduce the dimensionless variables $a = N_A k_b T/\epsilon E_0^2$, $G = g/E_0 I_b$ and the symbol $'$ to indicate $\frac{\mathrm{d}}{\mathrm{d}\xi}$.

Equation 1.23 can be formally rendered explicit, distinguishing so the local (first term) and non-local contribution (terms with the spatial derivatives) [57]

$$Y = \frac{G}{Q} - a\frac{Q'}{Q} + \frac{GY'}{Q} + a\frac{Y''}{1 + Y'}, \tag{1.24}$$

we note that a nonlinearity similar to that occurring in Kerr media (Kerr-saturated) emerges when non-local terms are weak. In conditions where $l \gg x_q$ nonlocal effects play a minor role, the spatial derivatives scale as x_q/l and a is of the order of unity, $\eta = x_q/l$ represents a smallness parameter so that

$$Y^{(0)} = \frac{G}{Q} + o(\eta), \tag{1.25}$$

this solution can be iterated in Eq. 1.24 and reads

$$Y^{(1)} = \frac{G}{Q} - a\frac{Q'}{Q} - \frac{Q'}{Q}\left(\frac{G}{Q}\right)^2 + o(\eta), \qquad (1.26)$$

The first term, that is the dominant in biased conditions, gives the Kerr-saturated or screening nonlinearity at the basis of photorefractive solitons. It implies a decrease of the effective field E with respect to E_0 ($G \simeq 1$) as a result of charge rearrangements. The second term can be identified with the diffusive field, whereas the third emerges from its coupling with the saturation (screening) field. These fields produce an antisymmetric contribution to the space-charge field, even for symmetric initial conditions. This fact leads, in long propagation regimes, to the bent propagation observed in [59].

1.3.3 Electro-optic Response

The field E we have derived leads to a modulation of the index of refraction through the electro-optic effect that can be evaluated with the power series expansion in Eq. 1.1. However, in general, the elements of the impermeability tensor are very small, in order to observe any nonlinear effects a large electro-optic response of the material is required. This can be achieved in non-centrosymmetric phases, such as in poled ferroelectrics, and in centrosymmetric paraelectric phases in proximity of the ferroelectric phase transition.

We focus our attention on the case of centrosymmetric media, where the linear response vanishes due to the system symmetry. The response assumes a scalar form along the optical axis, so that g_{ijkl} is diagonal and the index of refraction variation $\Delta n(E)$ becomes

$$\Delta n(E) = -\frac{1}{2}n^3 g_{eff}\varepsilon_0^2(\varepsilon_r - 1)^2 E^2, \qquad (1.27)$$

using this expression in Eq. 1.25 we obtain the canonical Kerr-saturated nonlinearity

$$\Delta n(I) = -\Delta n_0 \frac{1}{(1 + I/I_b)^2}, \qquad (1.28)$$

with $\Delta n_0 = (1/2)n^3 g_{eff}\varepsilon_0^2(\varepsilon_r - 1)^2 E_0^2$. This nonlinearity, depending on the sign of g_{eff}, has a focusing or defocusing effect on the propagating beam for $\Delta n_0 > 0$ and $\Delta n_0 < 0$ respectively. The fact that the intensity appears only through the ratio I/I_b is a consequence of the cumulative response and is the basis for the low powers needed for nonlinear optics in photorefractive media.

1.4 Nonlinear Optical Propagation

To understand nonlinear wave dynamics and soliton formation, we show how an index change induced by photorefraction (Eq. 1.28) modifies beam propagation. We consider a monochromatic electromagnetic wave with frequency $\omega = 2\pi c/\lambda$

$$\boldsymbol{E}_{opt}(\boldsymbol{r}, t) = \boldsymbol{E}_{\omega}(\boldsymbol{r}, t)e^{i\omega t} + c.c. \tag{1.29}$$

From Maxwell's equations its propagation in a medium that is homogeneous on scales of the order of λ follows the linear differential equation (Helmholtz equation) [13]

$$\nabla^2 \boldsymbol{E}_{\omega} + k_0^2 n^2 \boldsymbol{E}_{\omega} = 0, \tag{1.30}$$

where $k_0 = \omega/c$ and n is the refractive index tensor, which depends on the spatial coordinates via the electro-optic effect $n = n(\boldsymbol{r}, \omega)$. Assuming that it can be expressed as a small perturbation to the linear index of refraction, $n(\boldsymbol{r}, \omega) = n_0(\omega) + \Delta n(\boldsymbol{r}, \omega)$ with $n_0(\omega) \gg \Delta n(\boldsymbol{r}, \omega)$, we have: $n^2(\boldsymbol{r}, \omega) = n_0^2(\omega) + 2n_0\Delta n(\boldsymbol{r}, \omega)$.

From the Helmholtz equation under the slow-varying amplitude approximation, that implies $\partial_{zz} A_{(\omega),i}(x, y, z) \approx 0$, for an electro-optic crystal in a paraelectric m3m phase [12] we obtain the paraxial wave equation

$$\left[\frac{\partial}{\partial z} + \frac{i}{2k}\nabla_{\perp}^2\right] \boldsymbol{A}_{(\omega)}(x, y, z) = -\frac{ik}{n_0}\overset{=}{\Delta n} : \boldsymbol{A}_{(\omega)}(x, y, z), \tag{1.31}$$

where

$$\overset{=}{\Delta n} = \begin{bmatrix} n_{11} & n_{12} \\ n_{12} & n_{22} \end{bmatrix}, \tag{1.32}$$

$$\begin{aligned} n_{11} &= -\frac{1}{2}n^3\varepsilon_0^2(\varepsilon_r - 1)^2(g_{11}E_x^2 + g_{12}E_y^2), \\ n_{22} &= -\frac{1}{2}n^3\varepsilon_0^2(\varepsilon_r - 1)^2(g_{12}E_x^2 + g_{11}E_y^2), \\ n_{12} &= -\frac{1}{2}n^3\varepsilon_0^2(\varepsilon_r - 1)^2(g_{44}E_xE_y). \end{aligned} \tag{1.33}$$

In isotropic media and for the one-dimensional case, we have

$$\left[\frac{\partial}{\partial z} + \frac{i}{2k}\frac{\partial^2}{\partial^2 x}\right] \boldsymbol{A}_{(\omega)}(x, z) = -\frac{ik}{n_0}\Delta n \boldsymbol{A}_{(\omega)}(x, z). \tag{1.34}$$

This equation is formally similar to the Schrödinger equation, and is known as generalized Nonlinear Schrödinger Equation (generalized NLSE) and describes paraxial nonlinear wave propagation in the spatial domain. The longitudinal length variable z plays the role of time and the nonlinear term $-\frac{ik}{n_0}\Delta n$ plays the role of an effective potential for the light beam.[1] An analogous equation also holds for temporal propagation [60]. The term $\frac{i}{2k}\frac{\partial^2}{\partial^2 x}$ represent light diffraction that can be balanced for specific values of the nonlinearity depending on the external parameters. This compensation leads to a solution A_ω that is spatially-localized and stationary: a spatial soliton. In particular, with a nonlinearity given by Eq. 1.28 we obtain photorefractive screening solitons.

1.5 Photorefractive Soliton

An optical spatial soliton is a localized light beam that does not suffer distortion due to diffraction. This confinement can be achieved in materials where the refractive index locally depends on the optical field. Steady-state photorefractive solitons[2] in the one-dimensional centrosymmetric case can be identified as non-evolving solution of Eq. 1.31 with Eq. 1.28. The soliton amplitude depends on the propagation coordinate z only through a phase factor so that, using a self-consistent method, we look for solutions of the form

$$A(x, z) = u(x)e^{i\Gamma z}\sqrt{I_b}, \tag{1.35}$$

where Γ is the propagation constant. We renormalize the spatial coordinate x according to the following definitions

$$\xi \equiv \frac{x}{d}, \qquad d \equiv (\pm 2kb)^{1/2}, \qquad b = \frac{k}{n}\left[\frac{1}{2}n^3 g_{eff}\varepsilon_0^2(\varepsilon_r - 1)^2\left(\frac{V}{L}\right)^2\right]. \tag{1.36}$$

The quantity d is the so-called nonlinear length and its sign reflects the focusing ($g_{eff} > 0$) or defocusing ($g_{eff} < 0$) character of the nonlinearity. From the paraxial equation we obtain the dimensionless nonlinear wave equation [63]

$$\frac{\mathrm{d}^2 u(\xi)}{\mathrm{d}\xi^2} = \pm\left[\frac{1}{1+u_0^2} - \frac{1}{\left(1+u(\xi)^2\right)^2}\right]u(\xi), \tag{1.37}$$

with the dimensionless intensity $u_0^2 = I/I_b$ ad the $\pm$ sign corresponding to that of $\Delta n(I)$ [64, 65]. We consider here the focusing case, which gives bright optical solitons. Since Eq. 1.37 is non-integrable, its solutions identifying specific soliton waveforms are found via numerical integration. These solutions represent an attractor for the optical dynamics and the input beam profile reshapes itself to excite them.

---

[1] The analogy works in the case of linear Δn, independent on light intensity, for solitons $-\frac{ik}{n_0}\Delta n$ is not an effective potential.

[2] Two conditions can be identified among photorefractive solitons: steady-state [61] where the localization is stationary or quasi-steady-state [62] where it is transient.

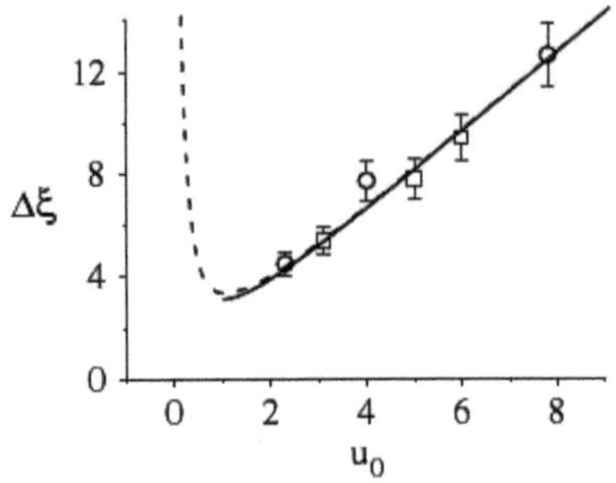

Fig. 1.3 Photorefractive soliton existence curve. Numerical prediction of Eq. 1.37 (dashed line), experimental results and analytical asymptotic result of segmented wave harmonic theory (black line). From [66]

In experiments, the accessible parameters are the nonlinear length b, the beam full-width at-half-maximum (FWHM) Δx and the peak intensity u_0^2. The parameter space $(u_0, \Delta\xi)$ is the nonlinear wave phase-space whose points lead to z-independent solutions define the soliton existence curve. The attractive nature of the existence points can thus be rephrased affirming that the soliton will have approximately the same u_0 and $\Delta\xi$ of the Gaussian input beam.

In Fig. 1.3 is reported the theoretical existence curve with experimental data in photorefractive potassium-lithium-tantalate-niobate (KLTN) crystals [67]. For $(u_0 > 1)$ the behavior can be approximated as $\Delta\xi = (\pi/2)(1+u_0^2)/u_0$, whereas in highly-saturated regimes $(u_0 \gg 1)$ segmented wave harmonic theory gives the asymptotic linear trait $\Delta\xi = (\pi/2)u_0$ [66]. This picture can be partially extended to the $(2+1)D$ (two-dimensional) case, where it becomes a three-dimensional nonlinear problem and assumes anisotropy and spatial non-locality. In this case, non-local contributions in the space-charge field are crucial to soliton existence, since the circular symmetry is broken by the tensorial nature of electro-optic response via the direction of the applied external field. The added spatial dimension implies a field E with components in both transverse dimensions according with Eq. 1.19, although the electro-optic index modulation maintains its scalar form.

The treatment can be extended to non-equilibrium conditions where the beam acquires a temporal dynamic and solitons can also be transient [68, 69]. From a phenomenological point of view, the initially diffracting light beam undergoes a cycle during which it first progressively self-focuses and settles into a self-trapped wave; then it undergoes a decelerated evolution during which the actual transverse beam intensity changes slightly, but the balancing of self-focusing and diffraction is approximately maintained; finally, it decays into a distorted and once again diffracting beam. To reveal the peculiar properties of the soliton state we have to reconsider the general case of Eq. 1.19. Introducing the so-called dielectric relaxation time $\tau_d = \frac{\varepsilon\gamma}{q\mu s\alpha I_b}$ so that the temporal variable is $\tau = t/\tau_d$, we can write the temporal-dependent counterpart of Eq. 1.25 as

$$\frac{\partial Y^{(0)}(\xi,\tau)}{\partial\tau} + Q(\xi,\tau)Y^{(0)}(\xi,\tau) = G. \tag{1.38}$$

Therefore, the dynamic space-charge field satisfies the integral equation

$$Y^{(0)} = Ge^{-\int_0^\tau Q\mathrm{d}\tau}\left[1 + \int_0^\tau \mathrm{d}\tau' e^{-\int_0^{\tau'} Q\mathrm{d}\tau''}\right]. \quad (1.39)$$

Although the model relies on a specific scale τ_d related to charge mobility, the dynamics described by Eq. 1.38 manifest several time scales. In particular, a stretched exponential behavior characterizes the process leading to both stationary (Steady-state) and non-stationary solitons (Quasi-steady-state) [70]. Time integration in Eq. 1.39 implies that Y at τ depends on Q at $\tau' < \tau$; the propagation has temporal nonlocality and allows memory effects. In proximity of the localization condition the normalized beam intensity Q becomes approximately time independent and, being $Q \ll 1$, Eq. 1.39 becomes

$$Y \simeq e^{Q\tau} + \frac{1}{Q} - \frac{1}{Q}e^{Q\tau} \rightarrow Y \simeq e^{Q\tau}. \quad (1.40)$$

The index variation associated to this space-charge field for a quadratic electro-optic response is the exponential nonlinearity $\Delta n = \Delta n_0 e^{-2Q\tau}$.

The approach of Eq. 1.35 can be generalizing with the spatiotemporal dimensionless variable $w(\xi) = \sqrt{2\tau}u(\xi)$ and applied to Quasi-steady-state solitons. The soliton profile equation reads [71].

$$\frac{\mathrm{d}^2 w(\xi)}{\mathrm{d}x^2} = -\left[\frac{1 - e^{-w_0^2}}{w_0^2} - e^{-w^2}\right] w(\xi), \quad (1.41)$$

where $w_0 = w(\xi = 0) = \sqrt{2\tau}u_0$. In Fig. 1.4a the generalized existence conditions in the variable $(w_0, \Delta\xi)$ are reported. This curve is accurate in conditions for which waveforms are approximately independent of time, or, in other word, in proximity of the minimum at w_0' corresponding to the onset of strong saturation in the nonlinearity, which also indicates a maximum value of nonlinear self-action. The minimum conditions in Eq. 1.41 gives a soliton width [71]

$$\Delta x_{min} = \frac{\Delta\xi_{min}\lambda}{2\pi n^2 \varepsilon\sqrt{g_{eff}}} E_0^{-1}. \quad (1.42)$$

An interesting case can also be derived. From Eq. 1.39 for an optical beam with random phase and amplitude variations on a fast spatial or temporal scale. Considering $Q = \bar{Q} + \Delta Q$, with fluctuations ΔQ having typical amplitude $\bar{Q}$ around zero, we found [58]

$$Y^{(0)} = Ge^{\bar{Q}\tau} + \frac{G}{\bar{Q}}(1 - e^{\bar{Q}\tau}), \quad (1.43)$$

that, for $\tau \gg 1/\bar{Q}$, reduces to the time-independent case $Y^{(0)} = G/\bar{Q}$. Therefore, the nonlinear response averages out fast intensity fluctuations leading to the steady

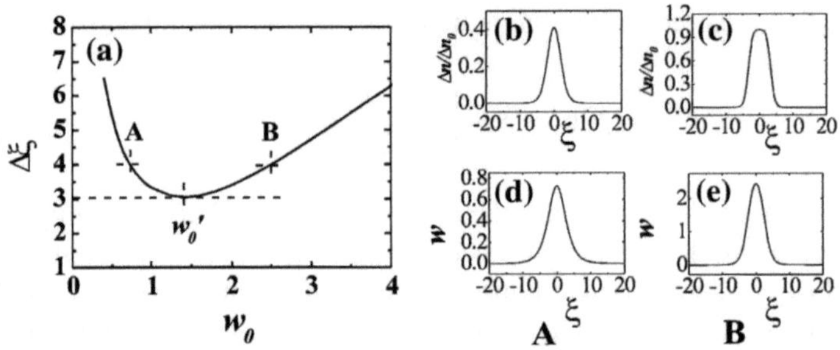

Fig. 1.4 **a** Existence curve for bright solitons of Eq. 1.41; **b, c** index patterns and **d, e** soliton profiles for the two points A and B, respectively, before and after the onset of strong saturation, highlighting the reshaping of the beam tails (**d** and **e**). From [71]

state condition as for a coherent beam. This is the basis for incoherent spatial solitons and it can be generalized to temporally incoherent (white) light [72, 73].

References

1. Brosi JM, Koos C, Andreani LC, Waldow M, Leuthold J, Freude W (2008) High-speed low-voltage electro-optic modulator with a polymer-infiltrated silicon photonic crystal waveguide. Opt Express 16(6):4177–4191
2. Ouskova E, Sio LD, Vergara R, White TJ, Tabiryan N, Bunning TJ (2014) Ultra-fast solid state electro-optical modulator based on liquid crystal polymer and liquid crystal composites. Appl Phys Lett 105(23):231122
3. Qianfan X, Schmidt B, Pradhan S, Lipson M (2005) Micrometre-scale silicon electro-optic modulator. Nature 435(7040):325–327
4. Roth M, Tseitlin M, Angert N (2005) Oxide crystals for electro-optic Q-switching of lasers. Glass Phys. Chem. 31:86–95
5. Malinowski A, Vu KT, Chen KK, Nilsson J, Jeong Y, Alam S, Lin D, Richardson DJ (2009) High power pulsed fiber MOPA system incorporating electro-optic modulator based adaptive pulse shaping. Opt Express 17(23):20927–20937
6. Goetz PG, Rabinovich WS, Mahon R, Murphy JL, Ferraro MS, Suite MR, Smith WR, Burris HR, Moore CI, Schultz WW et al (2012) Modulating retro-reflector lasercom systems for small unmanned vehicles. IEEE J Sel Areas Commun 30(5):986–992
7. Guarino A, Poberaj G, Rezzonico D, Degl'Innocenti R, Günter P (2007) Electro-optically tunable microring resonators in lithium niobate. Nat Photonics 1(7):407–410
8. Wang M, Yingxin X, Fang Z, Liao Y, Wang P, Chu W, Qiao L, Lin J, Fang W, Cheng Y (2017) On-chip electro-optic tuning of a lithium niobate microresonator with integrated in-plane microelectrodes. Opt Express 25(1):124–129
9. Goodby JW, Collings PJ, Kato T, Tschierske C, Gleeson HF, Raynes P (2014) Handbook of liquid crystals. Wiley-VCH, Weinheim
10. Ashkin A, Boyd GD, Dziedzic JM, Smith RG, Ballman AA, Levinstein JJ, Nassau K (1966) Optically-induced refractive index inhomogeneities in $LiNbO_3$ and $LiTaO_3$. *Appl Phys Lett* 9(1):72–74
11. Cavalieri AL, Fritz DM, Lee SH, Bucksbaum PH, Reis DA, Rudati J, Mills DM, Fuoss PH, Stephenson GB, Kao CC et al (2005) Clocking femtosecond X rays. Phys Rev Lett 94(11):114801

12. Yariv A, Yeh P (1984) Optical waves in crystals, vol 10. Wiley, New York
13. Born M, Wolf E (1980) Principles of optics: electromagnetic theory of propagation, interference and diffraction of light. Elsevier, Amsterdam
14. Aillerie M, Theofanous N, Fontana MD (2000) Measurement of the electro-optic coefficients: description and comparison of the experimental techniques. Appl Phys B 70(3):317–334
15. Park S-E, Shrout TR (1997a) Characteristics of relaxor-based piezoelectric single crystals for ultrasonic transducers. IEEE Trans Ultrason Ferroelectr Freq Control 44(5):1140–1147
16. Shrout TR, Park SE, Lopath PD, Meyer RJ, Ritter TA, Shung KK (1998) Innovations in piezoelectric materials for ultrasound transducers. In: Medical imaging 1998: ultrasonic transducer engineering, vol 3341. SPIE-International Society for Optical Engineering, pp 174–184
17. Uchino K (1996) Piezoelectric actuators and ultrasonic motors, vol 1. Springer Science & Business Media, Berlin
18. Dong WD, Finkel P, Amin A, Lynch CS (2012) Giant electro-mechanical energy conversion in [011] cut ferroelectric single crystals. Appl Phys Lett 100(4):042903
19. Booth ER, Wilbur ML (2004) Acoustic aspects of active-twist rotor control. J Am Helicopter Soc 1(49):3–10
20. Bokov AA, Ye Z-G (2006) Recent progress in relaxor ferroelectrics with perovskite structure. Frontiers of ferroelectricity. Springer, Berlin, pp 31–52
21. Shvartsman VV, Lupascu DC (2012) Lead-free relaxor ferroelectrics. J Am Ceram Soc 95(1):1–26
22. Ahart M, Somayazulu M, Cohen RE, Ganesh P, Dera P, Mao H-K, Hemley RJ, Ren Y, Liermann P, Wu Z (2008) Origin of morphotropic phase boundaries in ferroelectrics. Nature 451(7178):545–548
23. Lummen TTA, Gu Y, Wang J, Lei S, Xue F, Kumar A, Barnes AT, Barnes E, Denev S, Belianinov A et al (2014) Thermotropic phase boundaries in classic ferroelectrics. Nat Commun 5:3172
24. Borisevich AY, Eliseev EA, Morozovska AN, Cheng C-J, Lin J-Y, Chu Y-H, Kan D, Takeuchi I, Nagarajan V, Kalinin SV (2012) Atomic-scale evolution of modulated phases at the ferroelectric–antiferroelectric morphotropic phase boundary controlled by flexoelectric interaction. Nat Commun 3:775
25. Kutnjak Z, Petzelt J, Blinc R (2006) The giant electromechanical response in ferroelectric relaxors as a critical phenomenon. Nature 441(7096):956–959
26. Tian H, Meng X, Hu C, Tan P, Cao X, Shi G, Zhou Z, Zhang R (2016) Origin of giant piezoelectric effect in lead-free $K_{1-x}Na_xTa_{1-y}Nb_yO_3$ single crystals. Sci Rep 6
27. Park S-E, Shrout TR (1997b) Ultrahigh strain and piezoelectric behavior in relaxor based ferroelectric single crystals. J Appl Phys 82(4):1804–1811
28. Samara GA (2003) The relaxational properties of compositionally disordered ABO_3 perovskites. J Phys Condens Matter 15(9):R367
29. Lang SB, Chan HLW (2007) Frontiers of ferroelectricity: a special issue of the journal of materials science. Springer Science & Business Media, Berlin
30. Glinchuk MD, Eliseev EA, Morozovska AN (2008) Superparaelectric phase in the ensemble of noninteracting ferroelectric nanoparticles. Phys Rev B 78(13):134107
31. Rivera I, Kumar A, Ortega N, Katiyar RS, Lushnikov S (2009) Divide line between relaxor, diffused ferroelectric, ferroelectric and dielectric. Solid State Commun 149(3):172–176
32. Toulouse J, DiAntonio P, Vugmeister BE, Wang XM, Knauss LA (1992) Precursor effects and ferroelectric macroregions in $KTa_{1--x}Nb_xO_3$ and $K_{1-y}Li_yTaO_3$. Phys Rev Lett 68(2):232
33. Xu G, Zhong Z, Bing Y, Ye ZG, Shirane G (2006) Electric-field-induced redistribution of polar nano-regions in a relaxor ferroelectric. Nat Mater 5(2):134–140
34. Xu G, Wen J, Stock C, Gehring PM (2008) Phase instability induced by polar nanoregions in a relaxor ferroelectric system. Nat Mater 7(7):562–566
35. Akbarzadeh AR, Prosandeev S, Walter EJ, Al-Barakaty A, Bellaiche L (2012) Finite-temperature properties of Ba (Zr, Ti) O_3 relaxors from first principles. Phys Rev Lett 108(25):257601
36. Kleemann W (2014) Relaxor ferroelectrics: cluster glass ground state via random fields and random bonds. Phys Status Solidi B 251(10):1993–2002

37. Manley ME, Lynn JW, Abernathy DL, Specht ED, Delaire O, Bishop AR, Sahul R, Budai JD (2014) Phonon localization drives polar nanoregions in a relaxor ferroelectric. Nat Commun 5:3683
38. Phelan D, Stock C, Rodriguez-Rivera JA, Chi S, Leão J, Long X, Xie Y, Bokov AA, Ye ZG, Ganesh P et al (2014) Role of random electric fields in relaxors. Proc Natl Acad Sci 111(5):1754–1759
39. Pirc R, Kutnjak Z (2014) Electric-field dependent freezing in relaxor ferroelectrics. Phys Rev B 89(18):184110
40. Pirc R, Blinc R (1999) Spherical random-bond-random-field model of relaxor ferroelectrics. Phys Rev B 60(19):13470
41. Bokov AA, Ye Z-G (2012) Dielectric relaxation in relaxor ferroelectrics. J Adv Dielectr 2(02):1241010
42. Ishai PB, De Oliveira CEM, Ryabov Y, Feldman Y, Agranat AJ (2004) Glass-forming liquid kinetics manifested in a KTN: Cu crystal. Phys Rev B 70(13):132104
43. Ishai PB, Agranat AJ, Feldman Y (2006) Confinement kinetics in a KTN: Cu crystal: experiment and theory. Phys Rev B 73(10):104104
44. Viehland D, Jang SJ, Cross LE, Wuttig M (1990) Freezing of the polarization fluctuations in lead magnesium niobate relaxors. J Appl Phys 68(6):2916–2921
45. Wang S, Yi M, Bai-Xiang X (2016) A phase-field model of relaxor ferroelectrics based on random field theory. Int J Solids Struct 83:142–153
46. Pirc R, Blinc R (2007) Vogel-fulcher freezing in relaxor ferroelectrics. Phys Rev B 76(2):020101
47. Prosandeev S, Wang D, Akbarzadeh AR, Dkhil B, Bellaiche L (2013) Field-induced percolation of polar nanoregions in relaxor ferroelectrics. Phys Rev Lett 110(20):207601
48. Pugachev AM, Kovalevskii VI, Surovtsev NV, Kojima S, Prosandeev SA, Raevski IP, Raevskaya SI (2012) Broken local symmetry in paraelectric batio 3 proved by second harmonic generation. Phys Rev Lett 108(24):247601
49. Yokota H, Uesu Y, Malibert C, Kiat J-M (2007) Second-harmonic generation and x-ray diffraction studies of the pretransitional region and polar phase in relaxor $K_{(1--x)}Li_xTaO_3$. Phys Rev B 75(18):184113
50. Chang Y-C, Wang C, Yin S, Hoffman RC, Mott AG (2013a) Giant electro-optic effect in nanodisordered KTN crystals. Opt Lett 38(22):4574–4577
51. Chang Y-C, Wang C, Yin S, Hoffman RC, Mott AG (2013b) Kovacs effect enhanced broadband large field of view electro-optic modulators in nanodisordered KTN crystals. Opt Express 21(15):17760–17768
52. Gumennik A, Kurzweil-Segev Y, Agranat AJ (2011) Electrooptical effects in glass forming liquids of dipolar nano-clusters embedded in a paraelectric environment. Opt Mater Express 1(3):332–343
53. Yeh P (1993) Introduction to photorefractive nonlinear optics, vol 14. Wiley-Interscience, Hoboken
54. Boyd RW (2003) Nonlinear optics. Handbook of laser technology and applications (three-volume set). Taylor & Francis, Abingdon, pp 161–183
55. Kukhtarev NV, Markov VB, Odulov SG, Soskin MS, Vinetskii VL (1978) Holographic storage in electrooptic crystals. I. steady state. Ferroelectrics 22(1):949–960
56. Crosignani B, Di Porto P, Degasperis A, Segev M, Trillo S (1997) Three-dimensional optical beam propagation and solitons in photorefractive crystals. JOSA B 14(11):3078–3090
57. DelRe E, Ciattoni A, Crosignani B, Tamburrini M (1998a) Approach to space-charge field description in photorefractive crystals. JOSA B 15(5):1469–1475
58. DelRe E, Crosignani B, Di Porto P (2009) Photorefractive solitons and their underlying nonlocal physics. Prog Opt 53:153–200
59. DelRe E, D'Ercole A, Palange E (2005) Mechanisms supporting long propagation regimes of photorefractive solitons. Phys Rev E 71(3):036610
60. Agrawal GP (2007) Nonlinear fiber optics. Academic Press, Cambridge

61. Segev M, Valley GC, Crosignani B, Diporto P, Yariv A (1994) Steady-state spatial screening solitons in photorefractive materials with external applied field. Phys Rev Lett 73(24):3211
62. Duree GC Jr, Shultz JL, Salamo GJ, Segev M, Yariv A, Crosignani B, Di Porto D, Sharp EJ, Neurgaonkar RR (1993) Observation of self-trapping of an optical beam due to the photorefractive effect. Phys Rev Lett 71(4):533
63. Segev M, Agranat AJ (1997) Spatial solitons in centrosymmetric photorefractive media. Opt Lett 22(17):1299–1301
64. Chen Z, Garrett MH, Valley GC, Mitchell M, Shih M-F, Segev M (1996) Steady-state dark photorefractive screening solitons. Opt Lett 21(9):629–631
65. Wan W, Jia S, Fleischer JW (2007) Dispersive superfluid-like shock waves in nonlinear optics. Nat Phys 3(1):46–51
66. DelRe E, D'Ercole A, Agranat AJ (2003) Emergence of linear wave segments and predictable traits in saturated nonlinear media. Opt Lett 28(4):260–262
67. DelRe E, Crosignani B, Tamburrini M, Segev M, Mitchell M, Refaeli E, Agranat AJ (1998b). One-dimensional steady-state photorefractive spatial solitons in centrosymmetric paraelectric potassium lithium tantalate niobate. Opt Lett 23(6):421–423
68. Fressengeas N, Wolfersberger D, Maufoy J, Kugel G (1998) Build up mechanisms of (1+1)-dimensional photorefractive bright spatial quasi-steady-state and screening solitons. Opt Commun 145(1):393–400
69. Zozulya AA, Anderson DZ (1995) Nonstationary self-focusing in photorefractive media. Opt Lett 20(8):837–839
70. Dari-Salisburgo C, DelRe E, Palange E (2003) Molding and stretched evolution of optical solitons in cumulative nonlinearities. Phys Rev Lett 91(26):263903
71. DelRe E, Palange E (2006) Optical nonlinearity and existence conditions for quasi-steady-state photorefractive solitons. JOSA B 23(11):2323–2327
72. Christodoulides DN, Coskun TH, Mitchell M, Segev M (1997) Theory of incoherent self-focusing in biased photorefractive media. Phys Rev Lett 78(4):646
73. Mitchell M, Segev M, Coskun TH, Christodoulides DN (1997) Theory of self-trapped spatially incoherent light beams. Phys Rev Lett 79(25):4990

Chapter 2
Introduction to Microscopy

In this chapter we describe the basics and the state of the art of optical microscopy, the governing equation, the diffraction limit, image formation and the main criteria used for the analysis of imaging systems. In this framework, we also introduce the main concepts of super resolution, methods that can overcome the resolution limit using physical or numerical techniques, structured illumination microscopy, in which the sample is illuminated under different spatially inhomogeneous light fields and fluorescence microscopy, that images specific chemical components of a sample thanks to a selective staining using fluorophores.

2.1 Basics Concepts

A monochromatic field propagating in the increasing z direction in an homogeneous media can be decomposed as a sum of plane waves [1, Chap. 3.12]

$$\mathbf{E}(r) = \int_{\mathbf{k}_{||} \in \mathbb{R}^2} \mathbf{E}(\mathbf{k}_{||}) exp(i\mathbf{k} \cdot \mathbf{r}) d\mathbf{k}_{||}, \tag{2.1}$$

where $\mathbf{k}_{||} = k_x \mathbf{x} + k_y \mathbf{y}$ is the projection of $\mathbf{k}$ on the transverse (x, y) plane, $\mathbf{k}$ is the wave vector, with the constraint $||\mathbf{k}|| = k_0 = 2\pi/\lambda$ and $\mathbf{E}(\mathbf{k}_{||})$ verifies $\forall \mathbf{k}_{||}, \mathbf{E}(\mathbf{k}_{||}) \cdot \mathbf{k} = 0$. When $||\mathbf{k}|| \geqslant k_0$, kz is a complex number given by $k_z = i\sqrt{||\mathbf{k}||^2 - k_0^2}$ and $\mathbf{E}(\mathbf{k}_{||}) exp(i\mathbf{k} \cdot \mathbf{r})$ is an evanescent plane wave.

This plane wave decomposition is particularly useful for modeling image formation in optical microscopes. It is indeed natural to apply this decomposition using as z axis, the axis of symmetry of the imaging system, called the optical axis. The imaging system can then be modeled as a filter that collects some of these plane waves and transforms them in other plane waves reaching the image plane.

If the z-axis is oriented along the optical axis (the axis of symmetry of the optical instrument) and if this optical instrument has its aberrations corrected according to

G. Di Domenico, *Electro-optic Photonic Circuits*, Springer Theses,
https://doi.org/10.1007/978-3-030-23189-7_2

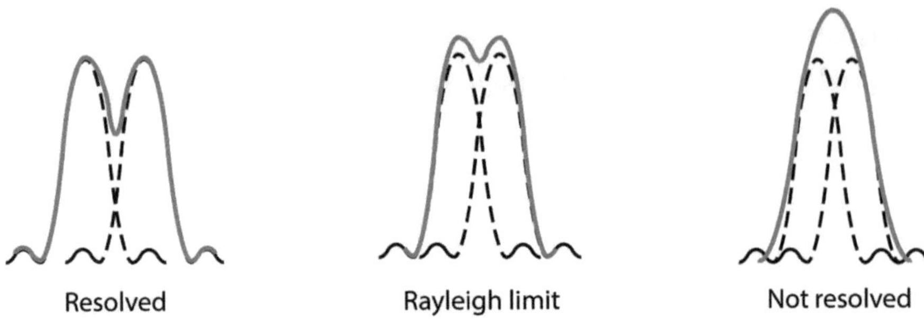

Fig. 2.1 Examples of Airy disks of 2 point sources in proximity of one another where (left) the sources are separated by a large distance (center) the Rayleigh limit where they are just resolvable (right) where they are unresolvable

the Sine-Abbe conditions [2], with its object focal plane at $z = 0$, plane waves such that $||k_{||}|| \leq k_0 NA$ are collected by this optical instrument and produce in the image plane the field

$$\mathbf{E}(r) = \int_{\mathbf{k}_{||} \in \mathbb{C}} \sqrt{\frac{k_z}{k_0}} \mathbf{E}(\mathbf{k}_{||}) exp(i\mathbf{k}' \cdot \mathbf{r}) d\mathbf{k}_{||}, \tag{2.2}$$

where $\mathbb{C}$ is the $k_0 NA$-radius disk centered in the origin of the (k_x, k_y) plane. NA is the numerical aperture, $NA = n_i sin(a)$ where a is the maximum polar angle of the plane waves collected by the objective and efficiently transformed by the imaging system. k', the wave vector in the image focal domain, in defined by

$$k' = \left(k_x/MF, k_y/MF, \sqrt{k_0^2 - (k_x/MF)^2 - \left(k_y/MF\right)^2}\right) \tag{2.3}$$

with $MF = n_i/n_d$ the magnifying factor [3, Chap. 4].

This decomposition in plane waves shows that the field frequencies above $k_0 NA$ are not collected by the optical system. It is this loss of information which explains the limit of resolution on the measurement of the field. However, microscopy is not interested in the field leaving the sample but rather in the sample itself. The main criteria describing the quality of the image formation process are Resolution, Contrast and Noise.

2.1.1 Resolution and Resolution Limit

Optical microscopes are ultimately limited by their resolution, the smallest distance at which two parallel lines can be resolved. According to Abbe, the limiting spot size resolution that can be achieved by a microscope is given by

$$r \geq \frac{\lambda}{2NA}. \tag{2.4}$$

The limit to resolution was redefined several times, primarily by Lord Rayleigh [4] and by the astronomer George Airy that had derived the Airy disk for high irradiance points of light [5]. Airy disks formed from the diffraction limited points of light show the best focused spots under conditions imposed by the optics and the nature of light. The disks display a central maximum of high irradiance surrounded by the subsequent dark and bright rings of the diffraction pattern of a point (see Fig. 2.1) whose positions are dependent on the wavelength of light and the numerical aperture or aperture size of the lens. The relation between the Airy disks and the resolving ability of the aperture can be shown with two sources shown by the Fig. 2.1. Each of these sources have an Airy disk associated with itself which spread over an angular distance of $\Delta\theta$ which is dependent on the equation:

$$\Delta\phi_{min} = \Delta\theta = \frac{1.22\lambda}{D} \tag{2.5}$$

where D is the diameter of the aperture (this also relates to the diameter of the acceptance cone of an objective which is equivalent to twice the numerical aperture). Where two point sources are a distance apart $\Delta\phi$, where $\Delta\phi > \Delta\theta$, the sources are resolvable. As the separation distance of the two points is reduced the Airy disks begin to overlap, up to a point where the 2 sources can no longer be resolved as separate entities and thus appear to be one larger point source.

Resolution can be improved by using a shorter wavelength of light for observations or by using an objective with a higher NA. Note that NA is related to the refractive index of a material, resolution can be improved by increasing the refractive index of a lensing material, this principle exploit in solid and oil immersion lenses.

2.1.2 Contrast

The contrast of an image measures the ability to distinguish the sample from the surrounding background. Different definitions of this criterion can be found in the literature. The one proposed by Michelson [6], that is sufficiently general: $C = (I_M - I_m)/(I_M + I_m)$, where I_M is the maximum intensity and I_m is the minimum. Obviously, a value of C close to 1 describes a highly contrasted image while a value close to 0 measures a weakly contrasted image, probably useless for interpretation. The contrast mechanisms depends on the interactions between the illumination light and the sample, principally absorption and reflection, however a large part of the interesting samples are almost transparent. They provide weakly contrasted images when they are illuminated by light in a classical microscope. To improve the contrast two main ways have been developed along the years: Phase contrast microscopy and the use of marker molecules. Phase contrast microscopy consists in modifying the set-up to get contrast from the refractive index variations [7–9]. It requires the use of interference and thus specific set-up for illumination and detection of coherent light). The other way consists in filling the sample with markers that produces a sufficient contrast in images. One of the main advantage of marking is the possibility to select markers (with different spectral behaviour) that target chemically one or more specific components of the sample, allowing targets to be imaged simultaneously. The most widely used are fluorescent markers. They have the ability to absorb light at a wavelength and to emit at a different wavelength. By using spectral filters, it is possible to remove the illumination field, obtaining thus a contrast equal to 1, whatever the sample.

2.1.3 Noise

A final important criterion is the noise strength. Noise deteriorates the image and prevents the observation of the finest details. The criterion ordinarily used to estimate this issue is the Signal-to-Noise Ratio $SNR = \langle I \rangle / \sigma_b$ where $\langle I \rangle$ is the average signal and σ_b is the noise standard deviation. Noise is commonly modeled as a white Gaussian process that apply independently on each pixel of the camera (since it is the sum of several sources of noise: thermal signal, electronic amplification, data transmission between the camera and the computer and more). For low intensity become important the so called shot-noise. It is due to the statistical nature of light emission by matter. In fact the number of photons reaching a pixel camera follows in general a Poisson statistics and only when the number of photons is large enough it can be approximation by a Gaussian statistic. Clearly the SNR is high for high intensity and low for low intensity. Noise is thus, in most cases, a limiting factor in fluorescence microscopy that has good contrast but limited brightness.

2.2 Optical Transfer Function

Aside from resolution, contrast and noise, a general and quantitative criterion to evaluate the quality of a linear optical system is the Optical Transfer Function or equivalently the Point Spread function. The Optical Transfer Function provides a curve of contrast with respect to the spatial frequency. It allows, for example, to define the resolution criterion with respect to noise, a detail becomes detectable when its contrast is greater than the noise level. The resolution can then be defined as the period for which the contrast is equal to the inverse of the SNR [10]. In a linear optical system the image M is the convolution of the function of interest O (sample field) by a certain point-spread-function h

$$M = O \otimes h. \tag{2.6}$$

Basically, h represents the image of a point object which is assumed to be same, whatever the position of the object. The Transfer Function is the Fourier transform $\tilde{h}$ of this convolution function. The Fourier transform of M and O fullfills $\tilde{M} = \tilde{O}\tilde{h}$. $\tilde{h}$ is thus the filter applied to $\tilde{O}$ during the measurement process.

One of the main characteristics of a Transfer Function $\tilde{h}$ is its support, namely the region of the Fourier space where it is non-null. This defines the frequencies of O that are accessible in the measurements. In microscopy, $\tilde{h}$ is usually a low-pass filter and it is zero beyond a bounded Fourier domain about the 0 frequency. All the object frequency information inside this Fourier domain is transmitted to the image M. All the frequency information that are outside this Fourier domain is lost. The radius of the support of the transfer function is called the frequency cut-off.

2.3 Super-Resolution

A whole range of de-noising and deconvolution algorithms [11, 12], using the experimental Optical Transfer Function can improve the contrast or the signal to noise ratio of the image. Yet, the frequency cut-off due to the bounded support of the Transfer Function remains. To improve the resolution, it is necessary to introduce a priori knowledge on the sample. To explain the fundamental limit of these numerical super-resolution techniques, the notion of Degree of Freedom of an image has been derived for coherent absorption microscopy [13, 14] and then extended to other forms of microscopy [15, 16]. One decomposes the linear operator that links the object quantity of interest to the recorded images using a Singular Value Decomposition (SVD). Namely, one builds an orthogonal basis of functions in the sample domain (sample eigenvectors) whose images through the operator form an orthogonal basis of the image plane (image eigenvectors). With a correct normalization, only a few of the image eigenvectors are above the noise level. The number of detectable sample eigenvectors yields the Degree Of Freedom of the imaging system. Whatever the numerical treatment applied to the image, only the detectable sample eigenvectors can be recovered.

A specific attention has now to be drawn to the use of positivity a priori information. Indeed, the values taken by the object function being often bounded, this a priori information is used in many deconvolution algorithm. For example, the density of fluorescent markers is physically positive and the relative permittivity of dielectric media has its real part superior to 1 and its imaginary part positive. Including this a priori information improves clearly the visual aspect and the apparent resolution of the reconstructed image, but may lead to artefacts or disappearance of interesting details. Sementilli et al. [17] proposed a method to evaluate the frequency radius up to which the Fourier components of the sample are correctly retrieved using this a priori information. For large field-of-view and common noise level there is almost no amelioration of the frequency cut-off.

Last, one may consider more stringent a priori information. For example, one can assume that the sample consists in lines or tubes or presents a high level of sparsity.

In its most extreme version, the sample can be assumed to be constituted of isolated emitters. In this case, one can localise, detect and separate two of them even if they are a lot closer than the resolution limit [18–20]. However, whatever the a priori information information added, one cannot decompose the sample images on a set of components larger than the Degree of Freedom [14, 21, 22]. The key point for high-resolution imaging is thus the development of technical solutions insuring that only few of these components contribute to each of the measurements. Promising recent super-resolution microscopy approaches (PALM, STORM in particular) are based on this idea.

2.4 Structured Illumination

An interesting approach to increase the spatial resolution of optical microscopy is to apply a patterned illumination field to the sample. In this approach, the spatial frequencies of an inhomogeneous illumination pattern mix with those of the sample features, shifting the high-frequency features to lower frequencies that are detectable by the microscope. Periodic illumination patterns can be created through the interference of multiple light sources in the axial direction [23], the lateral direction [24], or both [25]. By acquiring multiple images with illumination patterns of different phases and orientations, a high-resolution image can be reconstructed. Because the illumination pattern itself is also limited by he diffraction of light, structured illumination microscopy (SIM) is only capable of doubling the spatial resolution by combining two diffraction-limited sources of information. A resolution of ∼100 nm in the lateral direction and ∼300 nm in the axial direction has been achieved [23–26].

Noting P the illumination or probing function, (P depends on the chosen light-matter interaction, it corresponds to the incident field intensity in one-photon fluorescence microscopy and to the incident field in tomographic diffraction microscopy), and O the sample contrast distribution (which is either the fluorescence density or the relative permittivity respectively), the radiated signal is often proportional to the product OP. The imaging system does not act on the sample function O itself, but on this product, so that the recorded image is given by $M = (OP) \otimes h$. The convolution theorem states that the Fourier transform of a product is the convolution of the Fourier transforms,

$$\tilde{M} = (\tilde{O} \otimes \tilde{P})\tilde{h}. \tag{2.7}$$

Using an inhomogeneous probing P, information on some of the high frequencies of O are moved inside the measurements M. To separate the contributions of O and P in the filtered product, one takes several measurements with several probing fields: $M_n = (OP_n) \otimes h$. The sample contrast distribution O is then recovered from the many recorded images using a numerical treatment.

The oldest method using this structured illumination approach is scanning microscopy. The inhomogeneous illumination is a light spot obtained by focusing a wave into the smallest possible volume. This spot is then moved all over the sample.

Since $P(\mathbf{r}_0, \mathbf{r}) = P(\mathbf{r}_0 - \mathbf{r})$ is the illumination function produced in r when focusing on $\mathbf{r}_0$, the field in the image space is

$$M(\mathbf{r}_0, \mathbf{r}) = [O(\mathbf{r})P(\mathbf{r}_0 - \mathbf{r})] \otimes h(\mathbf{r}). \tag{2.8}$$

One can easily show that

$$M(\mathbf{r}_0, \mathbf{r}_0) = O(\mathbf{r}_0) \otimes [h(\mathbf{r})P(\mathbf{r}_0)] . \tag{2.9}$$

Equation 2.9 is the at the basis of confocal microscopy. This microscopy technique proposes to focus a laser beam, thanks to a microscope objective, on a point r_0 of the sample, and to detect through the same objective the light radiated by the same point. This technique exhibits an effective point-spread-function $h_{eff} = hP$ and an effective Transfer Function $\tilde{h}_{eff} = \tilde{h} \otimes \tilde{P}$. Further improvements can be obtained by collecting the whole light information of Eq. 2.8 [27–29] or by diminishing the size of the probing function P, in shaping the incident beam [30, 31], using non-linear contrast mechanisms [32], or near-field evanescent waves at the surface of nano-structured substrates [19, 33–36].

A second important implementation of the structured illumination principle is the inhomogeneous probing functions. The sample is not scanned by a spot but illuminated successively under many different illumination patterns P_n. This wide-field approach requires a numerical treatment of the different images to extract a correct estimation of O. The most classical illumination pattern is sinusoidal [24, 37]: $P_n(\mathbf{r}) = 1 + cos(\mathbf{K} \cdot \mathbf{r} + \phi_n)$, where $\mathbf{K}$ is the vector of the sinusoidal pattern and ϕ_n is a phase that has to be different for each illumination. This pattern is usually obtained via the interference of two coherent collimated beams. In this case, the Fourier transform of the images $M_n = (OP_n) \otimes h$ fullfills

$$\begin{aligned}
\tilde{M}_n(\mathbf{k}) &= \left[\tilde{O}(\mathbf{k}) \otimes \left(\delta(\mathbf{k}) + \frac{1}{2}exp(i\phi_n)\delta(\mathbf{k}+\mathbf{K}) + \frac{1}{2}exp(-i\phi_n)\delta(\mathbf{k}-\mathbf{K})\right)\right]\tilde{h}(\mathbf{k}) \\
&= \left(\tilde{O}(\mathbf{k}) + \frac{1}{2}exp(i\phi_n)\tilde{O}(\mathbf{k}+\mathbf{K}) + \frac{1}{2}exp(-i\phi_n)\tilde{O}(\mathbf{k}-\mathbf{K})\right)\tilde{h}(\mathbf{k}) \\
&= \tilde{M}_n^0(\mathbf{k}) + exp(i\phi_n)\tilde{M}_n^+(\mathbf{k}) + exp(-i\phi_n)\tilde{M}_n^-(\mathbf{k})
\end{aligned} \tag{2.10}$$

It is clear that $\tilde{M}_n^0(\mathbf{k}) = \tilde{O}(\mathbf{k})\tilde{h}(\mathbf{k})$ is the image that would be obtained under an homogeneous illumination and contains only the low frequencies of the object. On the contrary $\tilde{M}_n^{\pm}(\mathbf{k}) = \tilde{O}(\mathbf{k} \pm \mathbf{K})\tilde{h}(\mathbf{k})$ contains frequencies of O around the $\pm\mathbf{K}$ frequency. Using three different ϕ_n it is possible to separate these three components and to reconstruct $\tilde{O}$ in a Fourier domain that is larger than the support of $\tilde{h}$. The resulting Transfer Function depends on h and on $\mathbf{K}$. The same process can be repeated for different orientations of $\mathbf{K}$ in order to get an isotropic improvement.

2.5 Fluorescence

When specimens absorb and subsequently radiate light, we describe the process as photoluminescence. If the light emission persists for up to a few seconds after the excitation, the phenomenon is known as phosphorescence. Fluorescence, describes an optical process in which the molecular absorption of a photon leads to the excitation of an electron to a higher energetic state followed by relaxation to the ground state accompanied by the emission of another photon with a longer wavelength.

The use of the fluorescence microscope in research and lab based applications has increased significantly alongside the development of better fluorescent markers and the new techniques which enable more comprehensive exploitation of light for biologic imaging. These advances include the wide-spread use of fluorescent proteins [38] and the design of new fluorophores [39, 40] the development of wide-field fluorescence microscopy (WFFM) Techniques [41], The laser scanning confocal microscope (LSCM) [42, 43], the two-photon fluorescence microscopy (TPFM) [44, 45], the stimulated emission depletion fluorescence microscopy (STED) [46, 47].

Since excitation and emission occur at different wavelengths, it is possible to filter out the excitation light which ensures a nearly perfect contrast. Moreover, it is possible to fix these markers to specific targets and thus to image a specific chemical component of the sample. When the excitation intensity received by the fluorescent marker is low, the emitted intensity is proportional to the excitation intensity surrounding it [48]: $I_{out} = \sigma I_{ext}$, where I_{out} is the intensity emitted by the marker and I_{ext} is the intensity of the field at the absorption wavelength and at the position of the marker. The coefficient σ expresses the efficiency of the marker. Ordinary fluorescent markers are a lot smaller than the resolution reachable in optical microscopy. It is generally possible to consider a collection of markers diluted in a sample as a continuous density. One defines the function ρ of the space position r such that $\rho(r)dr = \sum_{l=1}^{L} \sigma_l$, where L is the number of marker in the small volume dr and σ_l is the emission coefficient of the lth fluorophore. However this modelling neglects three issues. The first is bleaching that is the coefficient σ actually decreases with time. More precisely, its decay is proportional to its emitted energy I_{out} [49]. the second is Blinking, besides this decay, there is a quick fluctuation of σ versus time. This fluctuation is useful for techniques like photoactivated localization microscopy (PALM) [50, 51] and and stochastic optical reconstruction microscopy (STORM) [52] and is at the basis of super-resolution optical fluctuation imaging (SOFI) [53]. The third issue is the near-field interactions: The coefficient σ of a fluorophore is actually influenced by its surroundings [54, 55]. For example, the emission of a single marker close to a mirror depends on its position with respect to the mirror.

Optimal use of fluorescence microscopy requires a basic understanding of the strengths and weaknesses of the various techniques as well an understanding of the fundamental trade-offs of the variables associated with fluorescent light collection [56].

2.6 Turbid Media

Imaging an object embedded in turbid media tissue is a fundamental topic for microscopy. Due to the multiple scattering effect in a turbid medium Fig. 2.2, the standard microscopic imaging theory, based on diffraction, is not necessarily applicable. Turbid media always exhibits complex characteristics as it has complex structures and is composed of various components. Usually, it shows a multiple-layer structure rather than a single-layer structure and consists of multiple sizes of scatterers rather than a single size. Furthermore, an inhomogeneous feature may exist because of the aggregation effect of the scatterers. Such structural, size, or aggregation features from turbid media will greatly influence images under a microscope. Research work in this field can be classified into two categories: transillumination imaging, in which case a parallel beam probe is used [57–61] and microscopic imaging, in which a microscopic objective is used for illumination [62–67]. An object embedded in a turbid medium is illuminated by an objective lens of a range of the illumination angle. The optical signal in each direction is made of two parts; the light scattered by the embedded object, the wanted signal, and by the scattering medium surrounding the object, an unwanted noise. Even if statistical analysis of scattered photon distributions shows that scattered photons still carry information about embedded objects [65, 68], the result of using an illumination objective (the two parts of the signal superpose each other) degrades the image quality.

A number of approaches have been proposed to obtain useful images through significant depths of a turbid medium. The currently available methods to selectively suppress the scattered photons based on the properties (denominated gating methods) are time-gating [69], which relies on the utilization of an ultrashort pulsed beam, coherence-gating [70], which relies on the degree of coherence of photons, polarization-gating [58], which relies on the polarization-state of photons, and angle-gating [71, 72], which relies on the path deviation of the scattered photons. Although all of these gating mechanisms can be employed in any imaging system, the efficiency of these methods depends on a particular imaging system. Transillumination imaging systems which use a parallel beam probe can give images of millimeter resolution

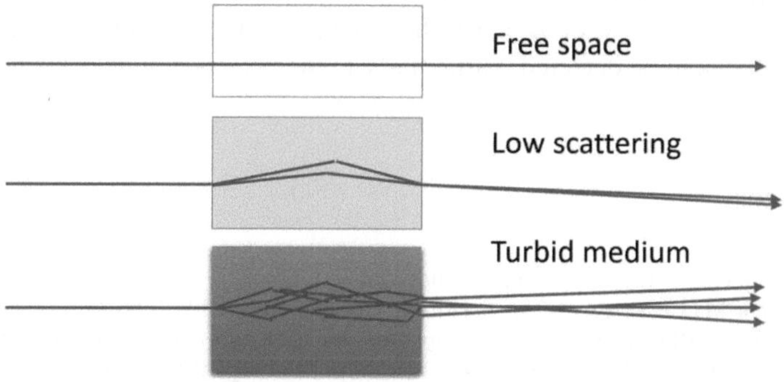

Fig. 2.2 Examples of microscopic scattering through free space (no scattering medium), a low scattering medium and a turbid medium

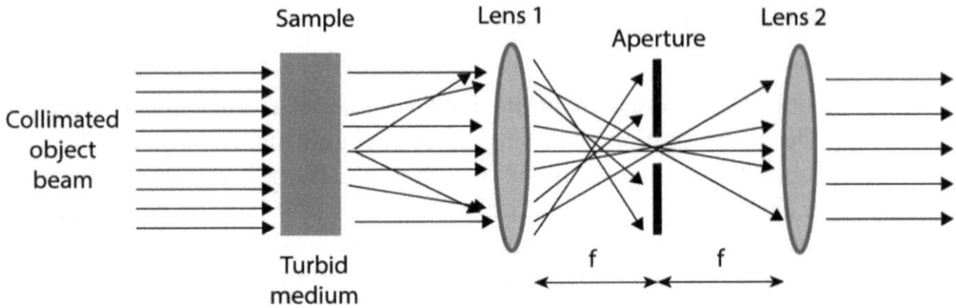

Fig. 2.3 Schematic diagram of a collimated beam propagating through a turbid medium and spatial filter

[73, 73]. To obtain an image of micrometer resolution, a microscope objective is necessary. In this case, time-gating may become less efficient due to the large range of illumination angles. However, angle-gating, polarization-gating, coherence gating, and fluorescence-gating are important in microscopic imaging.

In addition, the use of an objective leads to a focal region of an intensity that is high enough to produce nonlinear excitation such as two-photon excitation [74]. Because the strength of the nonlinear signal is mainly determined by the ballistic photons, any nonlinear excitation process under a microscope can be used to suppress scattered photons, which results in a unique nonlinear-gating mechanism in microscope imaging through turbid media [75].

However, using an objective lens in a microscopic imaging system raises the question of which numerical aperture of an objective is suitable for imaging [65]. According to the imaging theory based on Born's approximation, which ignores the multiple scattering in a turbid medium, a high numerical aperture objective lens can provide high diffraction-limited resolution [76–78]. Born's approximation is applicable to the case in which the optical thickness n, defined as the thickness of a turbid medium divided by the scattering mean free path length, is less than one. On the other hand, a low numerical aperture objective can suppress scattered photons that statistically travel at high angles (see Fig. 2.3). Both arguments are based on the assumption that ballistic light is dominant in forming an image. When a turbid medium is thick, e.g., when $n > 10$, the strength of the unscattered light/photons may be too weak to be detected, particularly in the presence of detector noise.

References

1. Jackson JD (1975) Electrodynamics. Wiley Online Library
2. Mansuripur M (1998) Abbe's sine condition. Opt Photon News 9(2):56–60
3. Alberto D (2001) Confocal and two-photon microscopy: foundations, applications and advances. Wiley-VCH. ISBN 0-471-40920-0, https://doi.org/10.1117/1.601121
4. Rayleigh L (1879) Xxxi. investigations in optics, with special reference to the spectroscope. Philos Mag 8(49):261–274

5. George Biddell Airy (1835) On the diffraction of an object-glass with circular aperture. Trans Camb Phil Soc 5:283
6. Michelson AA (1995) Studies in optics. Courier Corporation
7. Georges N (1960) Interferential polarizing device for study of phase objects. US Patent 2,924,142 9 Feb 1960
8. Zernike F (1942) Phase contrast, a new method for the microscopic observation of transparent objects. Physica 9(7):686–698
9. Zernike F (1942) Phase contrast, a new method for the microscopic observation of transparent objects part ii. Physica 9(10):974IN1981IN3983–980982986
10. Stelzer EHK (1998) Contrast, resolution, pixelation, dynamic range and signal-to-noise ratio: fundamental limits to resolution in fluorescence light microscopy. J Microsc 189(1):15–24
11. Luisier F, Blu T, Unser M (2011) Image denoising in mixed poisson-gaussian noise. IEEE Trans Image Process 20(3):696–708
12. Swedlow JR (2007) Quantitative fluorescence microscopy and image deconvolution. Methods Cell Biol 81:447–465
13. De Micheli E, Viano GA (2009) Inverse optical imaging viewed as a backward channel communication problem. JOSA A 26(6):1393–1402
14. Di Toraldo Francia G (1969) Degrees of freedom of an image. JOSA 59(7):799–804
15. Pierri R, Soldovieri F (1998) On the information content of the radiated fields in the near zone over bounded domains. Inverse Prob 14(2):321
16. Piestun R, Miller DAB (2000) Electromagnetic degrees of freedom of an optical system. JOSA A 17(5):892–902
17. Sementilli PJ, Hunt BR, Nadar MS (1993) Analysis of the limit to superresolution in incoherent imaging. JOSA A 10(11):2265–2276
18. Gelles J, Schnapp BJ, Sheetz MP (1988) Tracking kinesin-driven movements with nanometre-scale precision. Nature 331(6155):450–453
19. Gur A, Fixler D, Micó V, Garcia J, Zalevsky Z (2010) Linear optics based nanoscopy. Opt Express 18(21):22222–22231
20. Heintzmann R (2007) Estimating missing information by maximum likelihood deconvolution. Micron 38(2):136–144
21. Harris JL (1964) Diffraction and resolving power. JOSA 54(7):931–936
22. Saleh B (1977) A priori information and the degrees of freedom of noisy images. JOSA 67(1):71–76
23. Bailey B, Farkas DL, Taylor DL, Lanni F (1993) Enhancement of axial resolution in fluorescence microscopy by standing-wave excitation. Nature 366(6450):44–48
24. Gustafsson MGL (2000) Surpassing the lateral resolution limit by a factor of two using structured illumination microscopy. J Microsc 198(2):82–87
25. Gustafsson MGL, Shao L, Carlton PM, Rachel Wang CJ, Golubovskaya IN, Zacheus Cande W, Agard DA, Sedat JW (2008) Three-dimensional resolution doubling in wide-field fluorescence microscopy by structured illumination. Biophys J, 94(12):4957–4970
26. Schermelleh L, Carlton PM, Haase S, Shao L, Winoto L, Kner P, Burke B, Cardoso MC, Agard DA, Gustafsson MGL et al (2008) Subdiffraction multicolor imaging of the nuclear periphery with 3d structured illumination microscopy. Science 320(5881):1332–1336
27. Müller CB, Enderlein J (2010) Image scanning microscopy. Phys Rev Lett 104(19):198101
28. Sandeau N, Wawrezinieck L, Ferrand P, Giovannini H, Rigneault H (2009) Increasing the lateral resolution of scanning microscopes by a factor of two using 2-image microscopy. J Eur Opt Soc Rapid Publ 4:09040
29. Wicker K, Heintzmann R (2007) Interferometric resolution improvement for confocal microscopes. Opt Express 15(19):12206–12216
30. Dorn R, Quabis S, Leuchs G (2003) Sharper focus for a radially polarized light beam. Phys Rev Lett 91(23):233901
31. Sheppard CJR, Hegedus ZS (1988) Axial behavior of pupil-plane filters. JOSA A 5(5):643–647
32. Fujita K, Kobayashi M, Kawano S, Yamanaka M, Kawata S (2007) High-resolution confocal microscopy by saturated excitation of fluorescence. Phys Rev Lett 99(22):228105

33. Lemoult F, Lerosey G, de Rosny J, Fink M (2010) Resonant metalenses for breaking the diffraction barrier. Phys Rev Lett 104(20):203901
34. Sentenac A, Chaumet PC (2008) Subdiffraction light focusing on a grating substrate. Phys Rev Lett 101(1):013901
35. Van Putten EG, Akbulut D, Bertolotti J, Vos WL, Lagendijk A, Mosk AP (2011) Scattering lens resolves sub-100 nm structures with visible light. Phys Rev Lett 106(19):193905
36. Wang Z, Guo W, Li L, Luk'yanchuk B, Khan A, Liu Z, Chen Z, Hong M (2011) Optical virtual imaging at 50 nm lateral resolution with a white-light nanoscope. Nat Commun 2:218
37. Heintzmann R, Cremer C (1999) Laterally modulated excitation microscopy: improvement of resolution by using a diffraction grating. In: Proceedings of SPIE, vol 3568, p 15
38. Shaner NC, Steinbach PA, Tsien RY (2005) A guide to choosing fluorescent proteins. Nat Methods 2(12):905–909
39. Eisenstein M (2006) Helping cells to tell a colorful tale. Nat Methods 3(8):647–655
40. Suzuki T, Matsuzaki T, Hagiwara H, Aoki T, Takata K (2007) Recent advances in fluorescent labeling techniques for fluorescence microscopy. Acta Histochem Cytoc 40(5):131–137
41. Coling D, Kachar B (2001) Theory and application of fluorescence microscopy. Curr Protoc Neurosci 2–1
42. Hibbs AR (2004) Confocal microscopy for biologists. Springer Science & Business Media, Springer
43. Pawley JB (2006) Fundamental limits in confocal microscopy. In: Handbook of biological confocal microscopy, pp. 20–42. Springer, Berlin
44. Diaspro A, Bianchini P, Vicidomini G, Faretta M, Ramoino P, Usai C (2006) Multi-photon excitation microscopy. Biomed. Eng. Online 5(1):36
45. Svoboda K, Yasuda R (2006) Principles of two-photon excitation microscopy and its applications to neuroscience. Neuron 50(6):823–839
46. Kellner RR, Baier CJ, Willig KI, Hell SW, Barrantes FJ (2007) Nanoscale organization of nicotinic acetylcholine receptors revealed by stimulated emission depletion microscopy. Neuroscience 144(1):135–143
47. Willig KI, Rizzoli SO, Westphal V, Jahn R, Hell SW (2006) Sted microscopy reveals that synaptotagmin remains clustered after synaptic vesicle exocytosis. Nature 440(7086):935–939
48. Strickler SJ, Berg RA (1962) Relationship between absorption intensity and fluorescence lifetime of molecules. J Chem Phys 37(4):814–822
49. Song L, Hennink EJ, Ted Young I, Tanke HJ (1995) Photobleaching kinetics of fluorescein in quantitative fluorescence microscopy. Biophys J 68(6):2588–2600
50. Betzig E, Patterson GH, Sougrat R, Lindwasser OW, Olenych S, Bonifacino JS, Davidson MW, Lippincott-Schwartz J, Hess HF (2006) Imaging intracellular fluorescent proteins at nanometer resolution. Science 313(5793):1642–1645
51. Hess ST, Girirajan TPK, Mason MD (2006) Ultra-high resolution imaging by fluorescence photoactivation localization microscopy. Biophys J 91(11):4258–4272
52. Rust MJ, Bates M, Zhuang X (2006) Sub-diffraction-limit imaging by stochastic optical reconstruction microscopy (storm). Nat Methods 3(10):793–796
53. Dertinger T, Colyer R, Iyer G, Weiss S, Enderlein J (2009) Fast, background-free, 3d super-resolution optical fluctuation imaging (sofi). Proc Natl Acad Sci 106(52):22287–22292
54. Chance RR, Prock A, Silbey R (1974) Lifetime of an emitting molecule near a partially reflecting surface. J Chem Phys 60(7):2744–2748
55. Rahmani A, Chaumet PC, de Fornel F, Girard C (1997) Field propagator of a dressed junction: fluorescence lifetime calculations in a confined geometry. Phys Rev A 56(4):3245
56. Pearson H (2007) The good, the bad and the ugly. Nature 447(7141):138–140
57. Demos SG, Alfano RR (1997) Optical polarization imaging. Appl Opt 36(1):150–155
58. Morgan SP, Khong MP, Somekh MG (1997) Effects of polarization state and scatterer concentration on optical imaging through scattering media. Appl Opt 36(7):1560–1565
59. Papaioannou DG, Baselmans JJM, van Gemert MJC et al (1995) Image quality in time-resolved transillumination of highly scattering media. Appl Opt 34(27):6144–6157

60. Wang QZ, Liang X, Wang L, Ho PP, Alfano RR (1995) Fourier spatial filter acts as a temporal gate for light propagating through a turbid medium. Opt Lett 20(13):1498–1500
61. Yadlowsky MJ, Schmitt JM, Bonner RF (1995) Multiple scattering in optical coherence microscopy. Appl Opt 34(25):5699–5707
62. Deng X, Min G (2003) Penetration depth of single-, two-, and three-photon fluorescence microscopic imaging through human cortex structures: monte carlo simulation. Appl Opt 42(16):3321–3329
63. Deng X, Gan X, Min G (2004) Effective mie scattering of a spherical fractal aggregate and its application in turbid media. Appl Opt 43(14):2925–2929
64. Gan X, Min G (2002) Microscopic image reconstruction through tissue-like turbid media. Opt Commun 207(1):149–154
65. Gan XS, Schilders SP, Gu M (1998) Image formation in turbid media under a microscope. JOSA A, 15(8):2052–2058
66. Gu M, Gan X, Xiaoyuan D (2015) Microscopic imaging through turbid media: Monte Carlo modeling and applications. Springer, Berlin
67. Schilders SP, Gan XS, Gu M (1998) Effect of scatterer size on microscopic imaging through turbid media based on differential polarisation-gating. Opt Commun 157(1):238–248
68. Dunn A, DiMarzio C (1996) Efficient computation of time-resolved transfer functions for imaging in turbid media. JOSA A 13(1):65–70
69. de Haller EB (1996) Time-resolved transillumination and optical tomography. J Biomed Opt 1(7)
70. De Boer JF, Milner TE, van Gemert MJC, Stuart Nelson J (1997) Two-dimensional birefringence imaging in biological tissue by polarization-sensitive optical coherence tomography. Opt Lett 22(12):934–936
71. Gan X, Schilders S, Gu M (1997) Combination of annular aperture and polarization gating methods for efficient microscopic imaging through a turbid medium: theoretical analysis. Microsc Microanal 3(6):495–503
72. Schilders SP, Gan XS, Gu M (1998) Efficient suppression of diffusing photons using polarising annular objectives for microscopic imaging through turbid media. Bioimaging 6(2):92–97
73. Tromberg B, Yodh A, Sevick E, Pine D (1997) Diffusing photons in turbid media: introduction to the feature. Appl Opt 36(1):9–9
74. Denk W, Strickler JH, Webb WW et al (1990) Two-photon laser scanning fluorescence microscopy. Science 248(4951):73–76
75. Gan XS, Gu M (2000) Fluorescence microscopic imaging through tissue-like turbid media. J Appl Phys 87(7):3214–3221
76. Gu M (1996) Principles of three-dimensional imaging in confocal microscopes. World Scientific, Singapore
77. Gu M (2000) Advanced optical imaging theory, vol 75. Springer Science & Business Media, Berlin
78. Wilson T, Sheppard C (1984) Theory and practice of scanning optical microscopy, vol 180. Academic Press, London

Chapter 3
Miniaturized Photogenerated Electro-optic Axicon Lens Gaussian-to-Bessel Beam Conversion

In the article '*Miniaturized photogenerated electro-optic axicon lens Gaussian-to-Bessel beam conversion*' published in Applied Optics **56**, 10 (2017), we experimentally demonstrate an electro-optic Gaussian-to-Bessel beam-converter miniaturized down to a $30 \times 30\,\mu$m pixel in a KLTN paraelectric crystal. The converter is based on the electro-optic activation of a photoinduced and reconfigurable volume axicon-lens achieved using a pre-written photorefractive funnel space-charge distribution. The transmitted light beam has a tunable depth of field that can be more than twice that of a conventional beam with the added feature of being self-healing.

3.1 Introduction

A standard pixel in an image is generated by electro-optically or mechanically switching on and off a micrometric spot of light. This spot will naturally diffract causing a projected image to lose its resolution already after fractions of a meter for commercial $15 \times 15\ \mu$m single element matrices [1]. Although diffractive spreading is minimum for laser light, where each single pixel can ultimately be made to generate diffraction-limited Gaussian beams, it simply can never be eliminated as long as spots are used [2, 3].

We here demonstrate a Bessel-pixel (see Fig. 3.1a), an electrically activated tapered fiber index of refraction pattern in a paraelectric crystal capable of converting on command a Gaussian beam into a Bessel beam. The converter is based on miniaturized tunable axicon-like structures [4–9] that can form a basic building block for a wholly alternative image projection technology in which each single pixel of an image is carried in space by non-diffracting Bessel beams [10, 11]. The technique allows a lateral $30 \times 30\ \mu$m miniaturization with a low-voltage nanosecond response scalable into arrays [12–19]. Bessel beams also find applications in light sheet microscopy [20], in optical trapping [21] in imaging in refractive index inhomogeneity [22] and to induce optical lattices in nonlinear media [23]. They attract

G. Di Domenico, *Electro-optic Photonic Circuits*, Springer Theses,
https://doi.org/10.1007/978-3-030-23189-7_3

growing interest, for example at terahertz [24] and microwave [25] wavelengths, or in innovative schemes, such as in an achromatic optical regime [26]. They have been also successfully studied for holographic optical data storage application [27].

3.2 Theory

Gaussian-to-Bessel conversion is achieved using an electro-optic volume index of refraction pattern Δn that is photoinduced in a writing stage. In distinction to conventional spatial soliton studies in photorefractive crystals, where self-focusing compensates diffraction [28], here we produce a shrinking tapered shape Fig. 3.1b [9] so as to mimic the phase modulation produced by an axicon lens. The pattern, a biomimetic reproduction of tapered retinal glial-cells [29], depends on the applied bias field E_r through the relationship [4, 30]

$$\Delta n(Y) = \Delta n_0 \cdot (1/Q + Y - 1)^2 \cdot \exp(-2\tau Q), \tag{3.1}$$

where $Y = E_r/E_W$, E_W is the field used to photoinduce the axicon, E_r is the field used to activate the axicon in the reading stage, $Q = 1 + I_W(x, y, z)/I_b$, $I_W(x, y, z)$ is the intensity distribution used to photoinduce the pattern, and I_b is the homogeneous background illumination. The response is fixed by the parameters $(\Delta n_0, \tau)$, where $\Delta n_0 = -(1/2)n_b^3\epsilon_0^2(\epsilon_r(T) - 1)^2 g_{11} E_W^2$, n_b is the unmodulated crystal index of refraction, ϵ_0 the vacuum permittivity, $\epsilon_r(T)$ the temperature dependent low-frequency dielectric constant, g_{11} the appropriate quadratic electro-optic coefficient, $\tau = t_e/t_d$, where t_e is the duration of photoinduction process and t_d is the dielectric relaxation time [31]. In the Fig. 3.1b we show the results of simulation based on the theories described above. Particularly the calculated index of refraction pattern Δn for the reversed bias $Y = -0.6$ and its effect on the intensity distribution I of an input Gaussian laser beam, for conditions that reflect the experiments described below. However, comparison to experimental results can only be performed for the transmitted light distribution (see below) as the beam inside the volume leads to negligible lateral scattering and is hence unobservable. The actual properties of the Gaussian to Bessel conversion are hence fixed by the photoinducing beam, Δn_0, τ, L_z and Y. Limits to the achievable beams are dictated by anisotropy in the response and charge diffusion that arise to disort the index of refraction pattern in conditions of strong diffraction for small writing input FWHM [32].

3.3 Experimental

The Bessel pixel is demonstrated with the setup illustrated in Fig. 3.1c. A Gaussian light beam from a He-Ne laser operating at $\lambda \simeq 633$ nm is expanded and appropriately attenuated, focused (input intensity Full-Width-at-Half-Maximum $FWHM =$

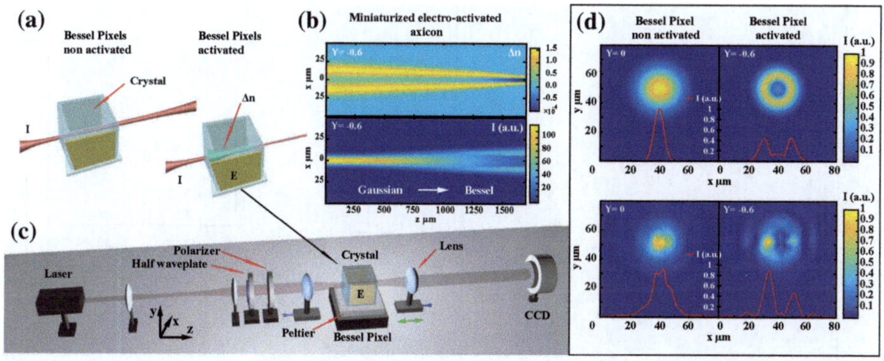

Fig. 3.1 **a** A schematic of light encoded by the Bessel-pixel. **b** Simulated Bessel-pixel refractive index pattern (numerical prediction from Eq. (3.1)) (top panel) and corresponding simulated Gaussian-to conical beam conversion (bottom panel). Electro-activation causes the funnel waveguide to anti-guide [8] and the Gaussian beam forms a donut-like diffraction. **c** Sketch of the experimental setup. **d** Comparison of the theoretical (top) and experimental (bottom) output intensity distribution for $Y = 0$ (OFF) and $Y = -0.6$ (ON). Observed diameter of the output donut is 20 μm and ratio between the peak intensity and intensity of the center for the beam (estimated neglecting the lateral lobe due to the bias electric field [32]) is 3.2. Note the lateral lobes (not present in the simulation) in the intensity pattern caused by the anisotropy in the space-charge distribution [32, 33]. Anisotropy can be reduced by reducing the exposure time and/or increasing the transverse size of the pattern

7.5 μm and power $P_r = 800$ nW) onto the input facet of a sample of $L_x = 1.8$ mm $\times$ $L_y = 2.2$ mm $\times$ $L_z = 1.7$ mm potassium-lithium-tantalate-niobate (KLTN) kept at $T = 22\,°\text{C}$ (above the sample Curie temperature $T_C = 16\,°\text{C}$). The beam is polarized parallel to the x axis and parallel to the applied external field E achieved delivering a constant voltage signal onto two silver electrodes on the x-facets, and pixel operation is monitored using a CCD. A closer view of the crystal is reported in Fig. 3.1a for both Bessel Pixel non activated (left) and activated (right), illustrating the focused laser beam I_R (red) and the funnel index pattern $\Delta n(Y)$ inside the sample (green). An imaging lens and a CCD camera is used to collect the beam intensity profiles, as shown in Fig. 3.1d for two paradigmatic cases that we term ON ($Y = -0.6$, $E_r = -1.7$kV/cm) and OFF ($Y = 0$, $E_r = 0$) states. In our setup, $n_b = 2.23$, $g_{11} = 0.14 m^2 C^4$, $E_W = 2.8$ kV/cm, $I_W(x, y, z)$ is the same He-Ne beam with higher intensity ($P = 10\,\mu$W) and focused onto the output facet of the sample heated to 32°C (to achieve the shape of Fig. 3.1b, $I_b \simeq 5 \cdot 10^3$ W/m^2 (20 times smaller than the input peak intensity of I_W), $t_e \simeq 390$ s and $t_d \simeq 180$ s. As reported in the profiles of Fig. 3.1d, the output intensity distributions for the two field conditions have regions of negligible overlap so that an appropriate mask can allow the pixel to act at once as a Bessel-beam generator and intensity modulation (not reported here).

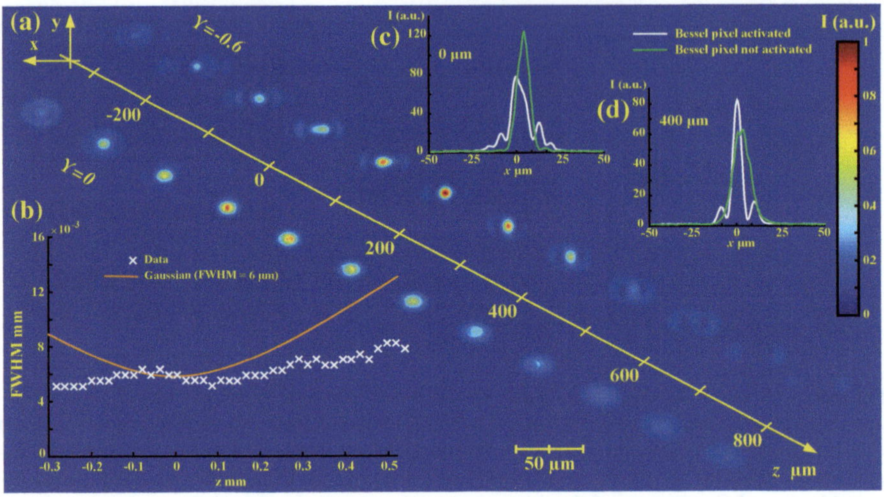

Fig. 3.2 Sequences **a** show the observed intensity distribution for the Bessel beam (corresponding to the ON state—top) and for the unmodulated Gaussian beam (OFF state—bottom) in different propagation planes. **b** Bessel beam FWHM (white crosses) compared to the theoretical Gaussian diffraction (orange line). We note that the use of the FWHM is meaningful only when the beam maintains a bell-like shape and breaks down for $z > 0.5$ mm. Also note that Bessel beam is compared to a theoretical Gaussian beam of 6 µm m because the Bessel beam has a central lobe that is smaller than the unmodulated Gaussian beam. Intensity profiles of the beam in the ON (white) and OFF (green) condition for **c** the $z = 0$ and **d** $z = 400$ µm plane

Evidence of Enhanced Depth of Field

To evaluate the depth of field, in Fig. 3.2a we compare the intensity distribution of the transmitted beams in the ON (top) and OFF (bottom) condition for different planes in proximity of the back focal plane. In the ON condition (Fig. 3.2a top), the pixel generates a central lobe with a FWHM of 6 µm that remains approximately constant even after 2.5 Rayleigh lengths, in sharp contrast to the diffraction of a corresponding 6 µm FWHM Gaussian solution (see the plot in Fig. 3.2b). In the OFF condition and without the annular mask enacting intensity modulation, the pixel generates a wider and diffracting 7.5 µm spot (Fig. 3.2a bottom).

Evidence of Self-healing

A second useful property of Bessel beams is their ability to recover their original shape and intensity when propagate through an obstacle [34], a feature commonly called 'self-healing'. To test the effect we compared beam transmission with and without a dust particle along its path 6 mm before the focal plane, deposited on an appropriate glass surface (see Fig. 3.3). When the funnel is deactivated so as to generate a standard Gaussian beam ($Y = 0$, bottom OFF condition of Fig. 3.2a),

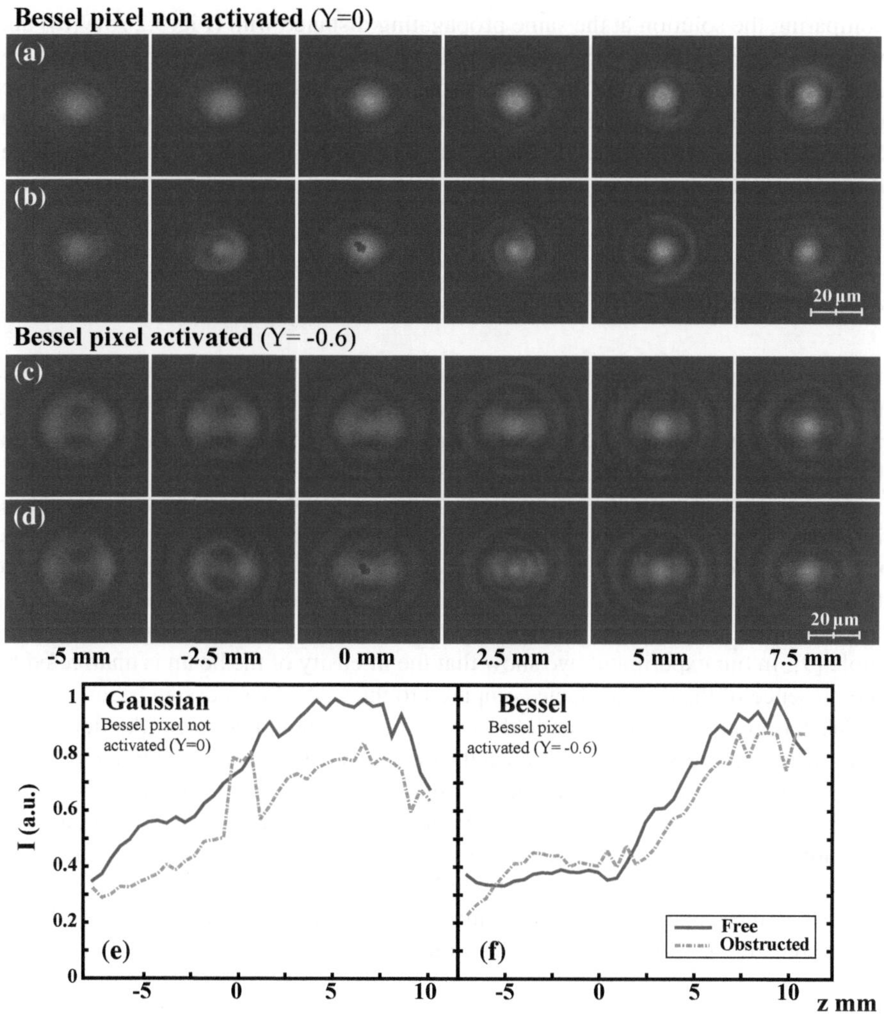

Fig. 3.3 Self-healing. **a, b, c** and **d** show, respectively, the intensity distribution of a Gaussian beam ($Y = 0$) for free propagation, for obstructed propagation, and the behavior of a Bessel beam ($Y = -0.6$) again for free and obstructed propagation. Distances are referred to the plane of the obstacle along the optical path. Plots (bottom) show the normalized peak intensity along the propagation **e** for a Gaussian beam ($Y = 0$) and **f** for a Bessel beam ($Y = -0.6$) both in the case of free (red line) and obstructed propagation (dashed blue line)

comparing the solution at the same propagating distance with (Fig. 3.2 row (b)) and without (Fig. 3.2 row (a)) the dust particle we can see that the obstructing particle distorts the beam, whereas in the case of an activated pixel ($Y = -0.6$, top ON condition of Fig. 3.2a), the Bessel-like pattern is rapidly recovered (see Fig. 3.3c, d)). Self-healing not only affects the shape but also for the intensity of the central lobe, that in the ON case is recovered (see Fig. 3.3e, f)).

3.4 Discussion

Finally, Bessel-Gauss beams are characterized by the maximum propagation distance, which varies with the focal of the Fourier lens [35]. In our study we found $z_{max} \simeq 900\,\mu\text{m}$ that is $\simeq$3.5 times the Rayleigh length of a Gaussian beam with a FWHM equal to that of the central lobe, see Fig. 3.2b.

We note that, in distinction to previous studies on self-focusing and self-defocusing in photorefractive and photogalvanic crystals [36], our approach does not involve self-action and beam nonlinearity, since when the beam causes the space-charge field the electro-optic effect is negligible, whereas when the electro-optic effect is made to affect propagation, the light-induced changes in the space-charge are negligible [9]. In our experiment, we show that the intensity of the beam is unaffected by the presence of the dust obstacle compared to the case of a Gaussian beam.

We have demonstrated a method to transform an input Gaussian-like field into a non-diffracting Bessel-like solution at the pixel level using a reconfigurable photoinduced axicon that allows a potentially fast and low-voltage volume integration into solid-state arrays. The effect involves a volume distributed conversion that, compared to conventional axicon-lens schemes, also avoids the unwanted scattering caused by manufacturing imperfections at the tip [37]. In our experiments, the beam generated possess all the properties of Bessel-Gauss beams, we find an increased depth-of-field and a signature self-healing property of the pixel beam.

References

1. Goodman JW, Steven C (1996) Introduction to fourier optics, vol 2. W H Freeman & Co (Sd), New York. https://doi.org/10.1117/1.601121 ISBN 0091-3286
2. Born M, Wolf E (1980) Principles of optics: electromagnetic theory of propagation, interference and diffraction of light. Elsevier, Amsterdam
3. Yariv A (1988) Quantum electronics. Wiley, Hoboken ISBN 0471609978
4. DelRe E, Pierangelo A, Palange E, Ciattoni A, Agranat AJ (2007) Beam shaping and effective guiding in the bulk of photorefractive crystals through linear beam dynamics. Appl Phys Lett 91(8):081105
5. DelRe E, Pierangelo A, Parravicini J, Gentilini S, Agranat AJ (2012) Funnel-based biomimetic volume optics. Opt Express 20(15):16631–16638

6. Parravicini J, Brambilla M, Columbo L, Prati F, Rizza C, Tissoni G, Agranat AJ, DelRe E (2014) Observation of electro-activated localized structures in broad area VCSELs. Opt Express 22(24):30225–30233
7. Parravicini J, Di Mei F, Pierangeli D, Agranat AJ, DelRe E (2015) Miniaturized electro-optic infrared beam-manipulator based on 3d photorefractive funnels. J Opt 17(5):055501
8. Pierangelo A, DelRe E, Ciattoni A, Palange E, Agranat AJ, Crosignani B (2008) Linear writing of waveguides in bulk photorefractive crystals through a two-step polarization sequence. J Opt A Pure Appl Opt 10(6):064005
9. Pierangelo A, Ciattoni A, Palange E, Agranat AJ, DelRe E (2009) Electro-activation and electro-morphing of photorefractive funnel waveguides. Opt Express 17(25):22659–22665
10. Durnin J, Miceli JJ Jr, Eberly JH (1987) Diffraction-free beams. Phys Rev Lett 58(15):1499
11. Gori F, Guattari G, Padovani C (1987) Bessel-gauss beams. Opt Commun 64(6):491–495
12. Asaro M, Sheldon M, Chen Z, Ostroverkhova O, Moerner WE (2005) Soliton-induced waveguides in an organic photorefractive glass. Opt Lett 30(5):519–521
13. Chauvet M, Guo A, Guoyan F, Salamo G (2006) Electrically switched photoinduced waveguide in unpoled strontium barium niobate. J Appl Phys 99(11):113107
14. DelRe E, Crosignani B, Di Porto P, Palange E, Agranat AJ (2002) Electro-optic beam manipulation through photorefractive needles. Opt Lett 27(24):2188–2190
15. DelRe E, Palange E, Agranat AJ (2004) Fiber-launched ultratight photorefractive solitons integrating fast soliton-based beam manipulation circuitry. J Appl Phys 95(7):3822–3824
16. D'Ercole A, Pierangelo A, Palange E, Ciattoni A, Agranat AJ, DelRe E (2008) Photorefractive solitons of arbitrary and controllable linear polarization determined by the local bias field. Opt Express 16(16):12002–12007
17. D'Ercole A, Palange E, DelRe E, Ciattoni A, Crosignani B, Agranat AJ (2004) Miniaturization and embedding of soliton-based electro-optically addressable photonic arrays. Appl Phys Lett 85(14):2679–2681
18. Fujihara T, Umegaki S, Hara M, Sassa T (2012) Formation speed and formation mechanism of self-written surface wave-based waveguides in photorefractive polymers. Opt Mater Express 2(6):849–855
19. Sapiens N, Weissbrod A, Agranat AJ (2009) Fast electroholographic switching. Opt Lett 34(3):353–355
20. Planchon TA, Gao L, Milkie DE, Davidson MW, Galbraith JA, Galbraith CG, Betzig E (2011) Rapid three-dimensional isotropic imaging of living cells using Bessel beam plane illumination. Nat Methods 8(5):417–423
21. Tsai YC, Leitz KH, Fardel R, Otto A, Schmidt M, Arnold CB (2012) Parallel optical trap assisted nanopatterning on rough surfaces. Nanotechnology 23(16):165304
22. Fahrbach FO, Simon P, Rohrbach A (2010) Microscopy with self-reconstructing beams. Nat Photonics 4(11):780–785
23. Rose P, Boguslawski M, Becker J, Diebel F, Denz C (2011) Light propagation in complex photonic lattices optically induced in nonlinear media. In: International workshop nonlinear photonics (NLP), 2011. IEEE, pp 1–3
24. Wei X, Liu C, Niu L, Zhang Z, Wang K, Yang Z, Liu J (2015) Generation of arbitrary order bessel beams via 3d printed axicons at the terahertz frequency range. Appl Opt 54(36):10641–10649
25. Cai BG, Li YB, Jiang WX, Cheng Q, Cui TJ (2015) Generation of spatial Bessel beams using holographic metasurface. Opt Express 23(6):7593–7601
26. Walde M, Jost A, Wicker K, Heintzmann R (2017) Engineering an achromatic bessel beam using a phase-only spatial light modulator and an iterative fourier transformation algorithm. Opt Commun 383:64–68
27. Manigo JP, Guerrero RA (2015) Self-imaging, self-healing beams generated by photorefractive volume holography. Opt Eng 54(10):104113–104113
28. DelRe E, Segev M (2009) Self-focusing and solitons in photorefractive media. Self-focusing: past and present. Springer, Berlin, pp 547–572

29. Labin AM, Ribak EN (2010) Retinal glial cells enhance human vision acuity. Phys Rev Lett 104(15):158102
30. DelRe E, Tamburrini M, Agranat AJ (2000) Soliton electro-optic effects in paraelectrics. Opt Lett 25(13):963–965
31. Gèunter P, Huignard J-P (2006) Photorefractive materials and their applications 1: basic effects. Springer, Berlin
32. DelRe E, Ciattoni A, Agranat AJ (2001) Anisotropic charge displacement supporting isolated photorefractive optical needles. Opt Lett 26(12):908–910
33. Korneev N, Marquez Aguilar PA, Sanchez Mondragon JJ, Stepanov S, Klein M, Wechsler B (1996) Anisotropy of steady-state two-dimensional lenses in photorefractive crystals with drift nonlinearity. J Mod Opt 43(2):311–321
34. Bouchal Z, Wagner J, Chlup M (1998) Self-reconstruction of a distorted nondiffracting beam. Opt Commun 151(4):207–211
35. Antonacci G, Di Domenico G, Silvestri S, DelRe E, Ruocco G (2017) Diffraction-free light droplets for axially-resolved volume imaging. Sci Rep 7(1):17
36. Song QW, Zhang CP, Talbot PJ (1993) Self-defocusing, self-focusing, and speckle in $LiNbO_3$ and $LiNbO_3$: Fe crystals. Appl Opt 32(35):7266–7271
37. Brzobohatỳ O, Čižmár T, Zemánek P (2008) High quality quasi-bessel beam generated by round-tip axicon. Opt Express 16(17):12688–12700

Chapter 4
Diffraction-Free Light Droplets for Axially-Resolved Volume Imaging

This chapter discusses the results of the article *'Diffraction-free light droplets for axially-resolved volume imaging'* published in Scientific Reports **7**, 17 (2017). We present non-diffracting light fields generated by the superposition of multiple Bessel beams, that produce an interference pattern along the propagation direction. We then focus our attention on how to use the effective finite depth of focus produced by the interference for fluorescent microscopy and for imaging in turbid media.

4.1 Introduction and Motivation

Light in inhomogeneous systems is strongly distorted because of the combined action of diffraction and scattering [1, 2]. The effects of diffraction can be arbitrarily reduced through nonlinearity [3, 4], to increase spatial resolution [5] and to allow light beams to penetrate deeper inside a turbid sample, but this at the expense of using specific materials in specific conditions. Airy beams [6, 7] and Bessel beams [8] are linear solutions of the Helmholtz Equation that are both non-diffracting and self-healing, features that make them naturally able to outdo the effects of penetration into a turbid volume. Pure Bessel beams result from the interference of a set of plane waves whose wave vectors form a cone [8], and their experimental realization, in the form of Bessel-Gauss beams [9] has been studied for material processing [10, 11], and optical tweezing [12, 13]. Their use in imaging was explored for light sheet microscopy [14], light field microscopy [15], for imaging in turbid systems [16], to enable imaging of live-cell dynamics [17], zebrafish embryos [18], and neuronal activity [19]. Ironically, it is this very non-diffracting nature with its enhanced penetration that makes them unsuitable for imaging in basic schemes, where their elongated dept of focus naturally washes out axial resolution along the propagation axis.

We demonstrate diffraction-free self-healing three-dimensional monochromatic light spots, what we term 'light droplet'. The droplets are able to penetrate deep into the volume of a sample, resist the effects of turbidity, and offer axial resolution comparable to that of Gaussian beams. The fields, formed from coherent mixtures of Bessel

G. Di Domenico, *Electro-optic Photonic Circuits*, Springer Theses,
https://doi.org/10.1007/978-3-030-23189-7_4

beams, have a discrete distribution of phase-velocities along the propagation axis. We show that light droplets manifest a more than ten-fold increase in their undistorted penetration, even in turbid milk solutions, compared to diffraction-limited beams. In a scanning fluorescence imaging scheme, we find a ten-fold increase in image contrast compared to diffraction-limited illuminations, and a constant axial resolution even after four Rayleigh lengths. Results pave the way to new opportunities in three-dimensional microscopy, where fast volume imaging can be achieved without the need of optical sectioning.

4.2 Experimental Methods

In our experiments, we achieve non-diffracting light fields with an effective finite depth of focus, comparable to that of a Gaussian beam. The apparently counterintuitive property emerges in the form of a non-diffracting spatial sequence of three-dimensional spots of light, that we term light droplets.

To describe the basic idea behind the droplets (more theoretical details in Sect. 4.3), we recall that a zeroth-order Bessel beam is formed by the monochromatic superposition of plane waves that propagate at a fixed angle θ with respect to a given axis $\hat{z}$. Since the waves share the same wavevector component $k_{\parallel} = k \cdot \cos\theta$, where $k = \sqrt{k_{\parallel}^2 + k_{\perp}^2} = 2\pi/\lambda$ and λ is the wavelength, there are no phase shifts along the axis between the interfering plane waves. This gives the beams a z-invariant filed distribution

$$E(\rho, z) = J_0(k_{\perp}\rho)\exp(jk_{\parallel}z), \tag{4.1}$$

where J_0 is the zeroth-order Bessel function and $\rho = \sqrt{x^2 + y^2}$ is the distance from the beam axis in the transverse x, y plane. To introduce z-dependence but preserve the diffraction-free and self-healing properties, we superimpose n coaxial Bessel beams, each with a distinct value of θ_i ($i = 1, ..., n$). The result is the interference light-droplet structure

$$I(\rho, z)_n = J_0(k_{\perp}\rho)^2 \left|\sum_{i=1}^{n} \exp\left(jk\cos\theta_i \cdot z\right)\right|^2. \tag{4.2}$$

Figure 4.1 shows a schematic of the principle behind our light droplets (Fig. 4.1a) and a diagram of the optical setup for a droplet structure with $n = 2$ (Fig. 4.1b). The laser beam (150 mW, 532 nm) was expanded at a magnification of 40× to illuminate a transmissive SLM (Holoeye, LC2012), which had a pixel pitch of 35 μm and a fill factor of 56%. To set the SLM in amplitude mode, the polarization of the incident laser was set at 45° to the $\hat{y}$ axis by a $\lambda/2$ wave-plate, and a crossed analyzer was placed after the SLM. A spherical lens of $f = 200$ mm focal length was placed next to the analyzer to perform a Fourier transform of the amplitude beam distribution. With

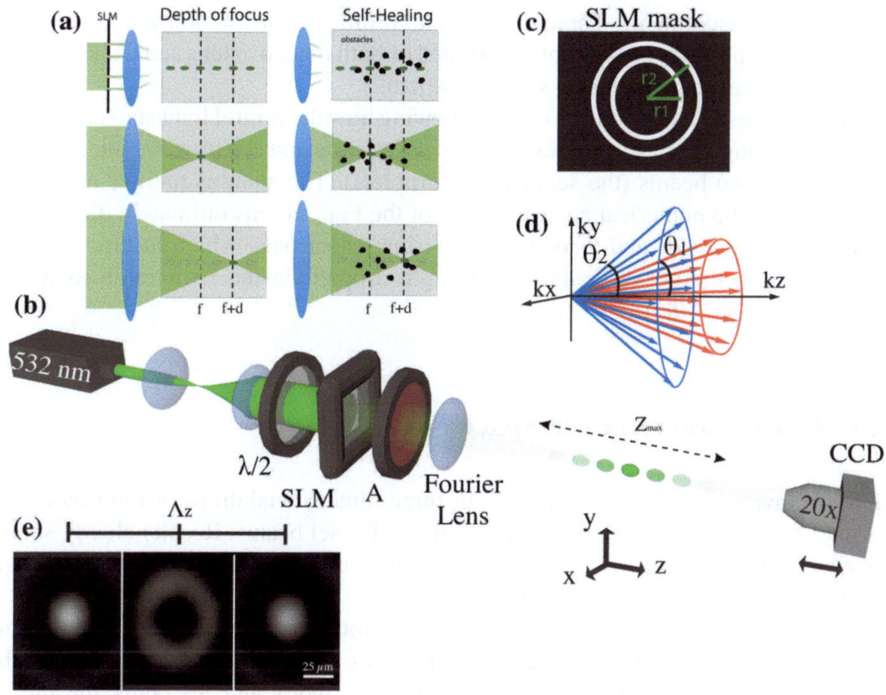

Fig. 4.1 Schematic of a light droplet illumination in contrast with a standard Gaussian configuration (**a**). A Gaussian beam can illuminate a single voxel at different depths in a sample volume through lenses of different focal lengths. In turn, light droplets are periodic structures which enable a localised illumination along the whole optical axis. Optical setup diagram (**b**). A binary mask (**c**) displaying two concentric annuli of different diameters is generated on the SLM. This operates in amplitude mode through a $\lambda/2$ plate and a crossed analyser (A). The resulting field distribution gives rise to two co-propagating Bessel beams of different k-vectors (**d**). The beams are recorded through a 20× objective lens and a CCD camera. Transverse intensity distributions at three different positions along a period of $\Lambda_z = 4\,\text{mm}$ (**e**)

a Gaussian beam illumination, this gave a minimum beam waist of $w_0 = 24.2\,\mu\text{m}$ and a Rayleigh length of $z_R = 3.5\,\text{mm}$. To obtain a single Bessel beam, a single ring of $r_1 = 1.7\,\text{mm}$ diameter and $150\,\mu\text{m}$ width was displayed on the SLM, whereas for two Bessel beams a second ring of $r_2 = 5.7\,\text{mm}$ diameter was added to the mask, as shown in Fig. 4.1c, these determining the corresponding θ_i and the associated k-vectors (Fig. 4.1d). The optimal beating between three Bessel beams was obtained by adding a third ring of $r_3 = 8.1\,\text{mm}$ diameter to the SLM mask. The droplet field is now generated using a spherical lens that performs a Fourier transform of the field distribution at the pupil plane where two co-propagating Bessel beams with $\theta_1 \neq \theta_2$ beat along the $\hat{z}$ axis.

All beams were imaged by a 20x objective lens (Olympus, UplanS) and a CCD camera (Thorlabs, DCC1545M) translating both along the $\hat{z}$ direction to maintain a constant magnification of the beam profile, with an equal data acquisition time in each set of measurements. Three dimensional profile reconstruction was performed by translating the imaging system constituted of both objective lens and CCD camera by a mechanical stage with step size of 200 μm.

Three planes are reported in Fig. 4.1e showing how the central light spot turns dark and bright along the $\hat{z}$ axis as a result of the destructive and constructive interference between the two beams (the sequence of droplets). The annulus radii r_1 and r_2 of the mask and the numerical aperture (NA) of the Fourier lens ultimately dictate the propagation angles θ_1 and θ_2 of the interfering wavefronts, which in turn sets the dimension and the period length of the light droplets in the $\hat{z}$ direction to $\Lambda_z = 2\lambda/[\cos(\theta_2) - \cos(\theta_1)]$.

4.3 Propagation in Free Space

Here we investigate the properties of our three-dimensional drops of light by comparing them experimentally with Gaussian and Bessel beams. Results clearly show the non-diffracting property of light droplets and their effective finite depth of focus, comparable to that of a Gaussian beam.

Figure 4.2 reports the beam intensity profiles along the $\hat{z}$ axis for a Gaussian beam and different light-droplet realizations. Distances are normalised to the Rayleigh length $z_R = \pi n w_0^2/\lambda$, with n being the refractive index and w_0 being the minimum beam waist of the Gaussian beam at the focal plane. In Fig. 4.2a the beams were obtained using a Gaussian mask, a single annulus ($n = 1$), two annuli ($n = 2$), and three concentric annuli ($n = 3$). Unlike the diffracting Gaussian beam, which diverges after the focal plane of the Fourier lens, the Bessel beam propagates along the optical axis with a constant width of the central peak. Diffraction-free light droplets arise from the interferometric beating of co-propagating Bessel beams for $n > 1$.

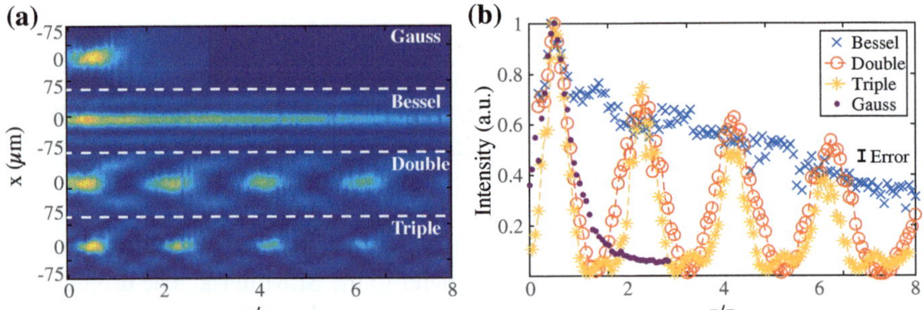

Fig. 4.2 Top view (**a**) and axial intensity profile (**b**) of Gaussian, Bessel beams ($n = 1$) and light droplets obtained with a double ($n = 2$) and a triple annular ($n = 3$) mask respectively. Whereas the Gaussian beam diverges after the lens focus, both Bessel beam and light droplets are shown to be non-diffracting

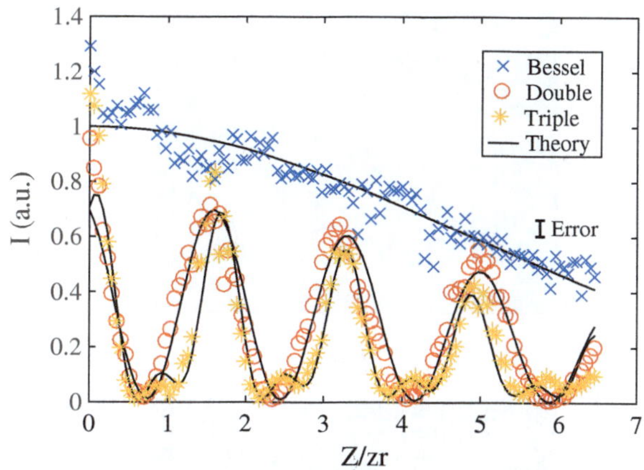

Fig. 4.3 Experimental data of the normalised beam intensity and their theoretical prediction (full curves) in the case of a simple Bessel-Gauss beam ($n = 1$) (blue circles), two Bessel-Gauss beams ($n = 2$) (red circles) and three Bessel-Gauss beams ($n = 3$) (green circles)

Figure 4.2b reports the x-y integrated intensity distribution of the beams along the $\hat{z}$ axis. Light droplets in the $n = 3$ case show a more localized profile in $\hat{z}$ compared to the $n = 2$ case, as expected from the three wave interference. For a fixed sample thickness and experimental scenario, a sufficiently high value of n will, in principle, give rise to a single diffraction-free light droplet.

Figure 4.3 illustrates the normalised peak intensity of a Gauss-Bessel beam and light droplets ($n = 2, 3$) measured along the $\hat{z}$ direction. The data shows good agreement with our theoretical predictions. In particular

$$E_i(\rho, z) = A\frac{w_0}{w(z)}\exp\left\{i\left[\left(k - \frac{k_\perp^2}{2k}\right)z - \Phi(z)\right]\right\}$$
$$\times J_0\left[\frac{\rho k_\perp}{\frac{iz}{z_R}+1}\right]\exp\left[\left(\frac{ik}{2R(z)} - \frac{1}{w^2(z)}\right)\left(\rho^2 + \frac{k_\perp^2 z^2}{k^2}\right)\right] \tag{4.3}$$

is the field distribution of a Bessel-Gauss beam [9], with A being an amplitude factor, $w(z) = w_0\sqrt{1 + (z/z_R)^2}$ the beam waist along the propagation, $R(z)$ the radius of curvature of the beam wavefronts and $\Phi = \tan^{-1}(z/z_R)$ the Gouy phase shift. Equation 4.3 generalizes both Gaussian and Bessel field solutions, where $E_1(r, z)$ reduces to a Gaussian field for $\theta = 0$, and to a Bessel field for $w_0 = 0$.

Experimental data in Fig. 4.3 was fitted using the intensity distribution given by $I(r, z) = |\sum_{i=1}^{n} E_i(r, z)|^2$, where $n = 1$ reduces to the standard case of a Bessel-Gauss beam, and $n = 2, 3$ gives solutions for the light droplets.

4.4 Droplet in Fluorescence Microscopy and Turbid Media

Most of today's imaging technologies rely on the use of fluorophore molecules to have high specificity and high contrast of the specimens against the background. In this section we experimentally demonstrate how droplet illumination can be used in

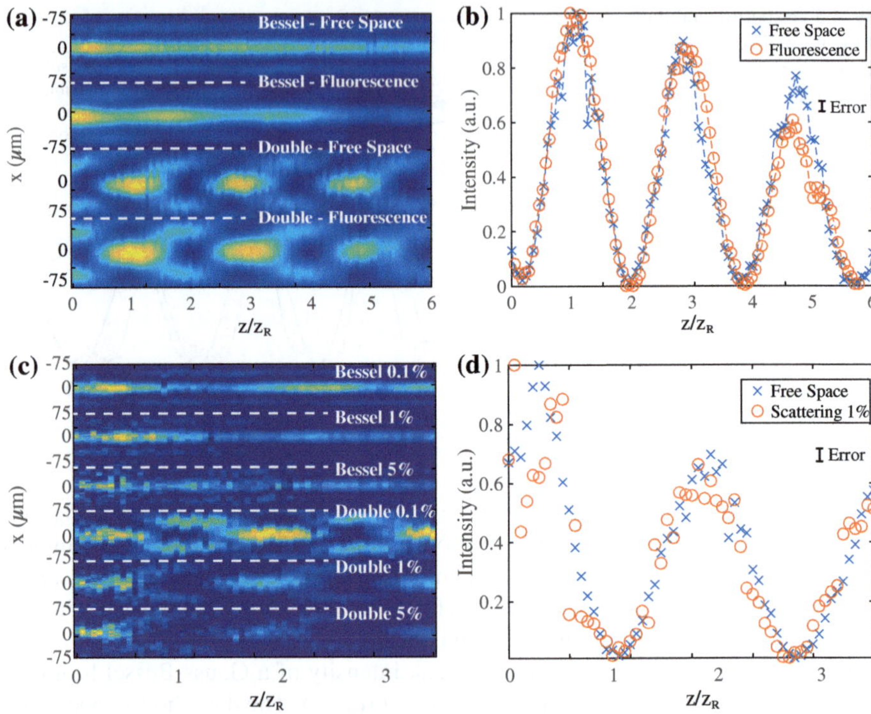

Fig. 4.4 Top view (**a**) and axial intensity profile (**b**) of the fluorescent signal given by the light droplet illumination. A thin glass layer labeled with *Rhod-2* fluorescent molecules was used as a sample. To maintain a constant magnification, the glass layer was translated along the $\hat{z}$ direction together with the imaging system. As expected, the fluorescent signal reproduces the illumination profile. Top view (**c**) and axial intensity profile (**d**) of light droplets in water-milk solutions at different concentrations. The beams are shown to be immune to scattering as result of their self-reconstruction property

combination of fluorophores and their ability to penetrate in the volume of biological specimens. Thus fulfilling basic requirement to generate three-dimensional images.

In Fig. 4.4a, b we demonstrate the use of light droplets to excite and detect fluorescent molecules typical of most imaging schemes. We measured the fluorescent emission given by the *Rhod-2* fluorescent dye labelling a thin glass layer. In Fig. 4.4a we report the resulting fluorescent signal given by both Bessel beam and light droplet illuminations along the optical axis. The labeled glass layer was first imaged at the CCD plane and then translated along the $\hat{z}$ direction together with both objective lens and CCD camera to maintain a constant magnification during the scan. A long pass filter was placed at the detector plane to reject all the incident light transmitted by the glass and to detect only the fluorescent signal emitted, which had a central wavelength at approximately 581 nm.

In Fig. 4.4c, d we demonstrate droplet self-healing in strongly scattering liquid volumes. We report the intensity distribution of light droplets compared to simple

Bessel beams in water-milk solution at concentrations that simulate real biological tissues (water-milk scattering is similar to that of intralipid solutions in the range of 1 and 10%) [20]. All beam profiles were taken along the propagation in a glass cuvette of 10 mm length filled with the scattering liquid. Despite an increasing noise level due to the dynamics of lipid droplets close the object plane, both Bessel beams and light droplets maintain equal shape and size as the milk concentration increases.

Droplet for Imaging System

In this last experiment we use light droplets in a standard confocal imaging system (Fig. 4.5). We report a set of images acquired in transmission mode with both Gaussian and droplet illuminations. In particular, a customized fluorescent sample was scanned in the $\hat{x}$ and $\hat{y}$ directions across the three different z-planes illustrated in Fig. 4.5a. As expected, both Gaussian and droplet images show equal contrast at the focal plane of the illumination lens ($z = z_1$), as shown in Fig. 4.5b. Images obtained

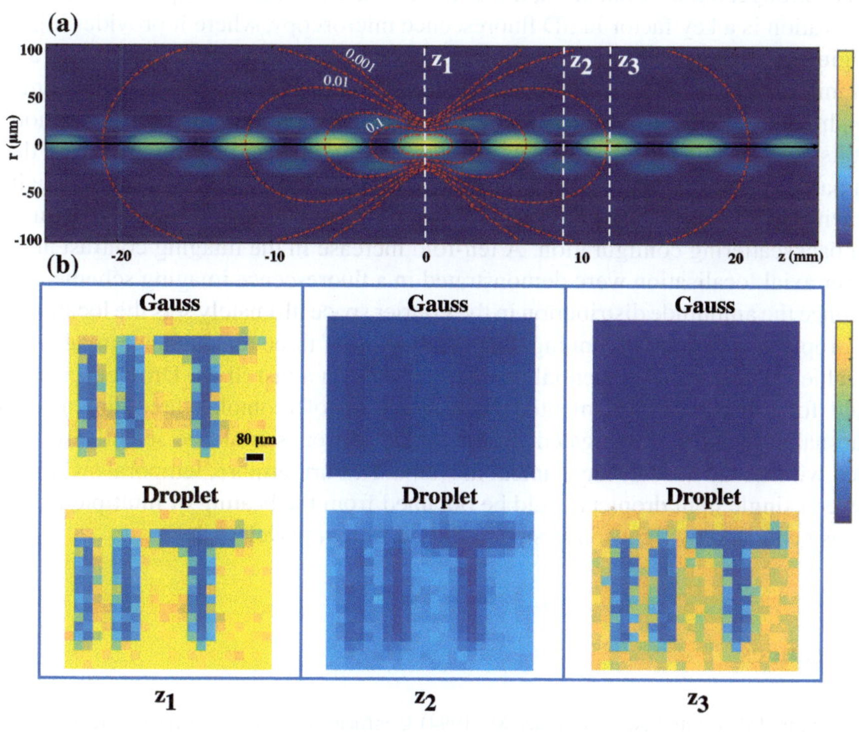

Fig. 4.5 Top view of the droplet intensity distribution and contour (red dashed) lines of a focused Gaussian beam (**a**). A set of images were acquired at three different z-positions along the illumination axis (**b**). Images acquired at the focal plane ($z = z_1$) with both Gauss and droplet illuminations exhibited equal contrast. However, as the sample was axially displaced from the focal plane, the contrast given by the Gaussian illumination rapidly dropped to zero, whereas the droplet illumination gave a decrease ($z = z_2$) followed by a 10×-increase ($z = z_3$) in the image contrast, the manifestation of the droplet axial-localization

with the Gaussian beam illumination exhibit the natural sharp diffraction-induced decrease in the imaging contrast as the sample is axially displaced from the illumination focal plane. In turn, the images acquired with the droplet illumination exhibited a significantly reduced contrast nearby the local droplet minimum ($z = z_2$) followed by a sharp increase at $z = z_3$, where the signal-to-noise-ratio was measured to be more than 10-fold higher than that given by the Gaussian illumination. Unlike Bessel beams, the finite depth of field of the droplets provides a high z-localization in deep regions of a sample.

4.5 Conclusion

Our results demonstrate that light droplets allow selective multi-point 3D localized illumination inside turbid volumes. Ideally, mixing more Bessel beams ($n > 3$) can lead to stronger axial localization, this at the expense of a more complex mask. Axial localization is a key factor in 3D fluorescence microscopy, where it provides higher imaging contrast given by a more efficient photon balance along the optical axis and minimizes background noise outside the excitation region, which in turn limits the photobleaching of the fluorescent molecules at a minimal level. Furthermore, results demonstrate that the droplets are self-healing in turbid liquids, a feature that can partially overcome diffusive scattering typical of biological tissues, potentially allowing *in vivo* imaging of thicker, more densely populated fluorescent specimens in a backscattering configuration. A ten-fold increase in the imaging contrast and a higher axial localisation were demonstrated in a fluorescence imaging scheme.

Since the amplitude distribution in the Fourier space ultimately sets the location of the droplets, motionless scanning data acquisition of three dimensional images can be achieved streaming sequential annular patterns on a fast SLM. Droplet scanning could, for example, be implemented when the activity of a complex three-dimensional neuronal network is investigated, where single neurones could be selectively activated without the need of mechanical movements. Furthermore, temporal switching between single light droplets could be obtained from the beating of multiple Bessel beams generated by light sources of slightly different wavelengths.

References

1. Schmitt JM, Knüttel A, Yadlowsky M (1994) Confocal microscopy in turbid media. JOSA A 11(8):2226–2235. https://doi.org/10.1364/JOSAA.11.002226. http://josaa.osa.org/abstract.cfm?URI=josaa-11-8-2226
2. Smithpeter CL, Dunn AK, Welch AJ, Richards-Kortum R (1998) Penetration depth limits of in vivo confocal reflectance imaging. Appl Opt 37(13):2749–2754
3. DelRe E, Spinozzi E, Agranat AJ, Conti C (2011) Scale-free optics and diffractionless waves in nanodisordered ferroelectrics. Nat Photonics 5(1):39–42

4. DelRe E, Di Mei F, Parravicini J, Parravicini G, Agranat AJ, Conti C (2015) Subwavelength anti-diffracting beams propagating over more than 1,000 rayleigh lengths. Nat Photonics
5. Barsi C, Fleischer JW (2013) Nonlinear abbe theory. Nat Photonics 7(8):639–643
6. Siviloglou GA, Broky J, Dogariu A, Christodoulides DN (2007) Observation of accelerating airy beams. Phys Rev Lett 99(21):213901
7. Zhang Peng, Yi Hu, Cannan Drake, Salandrino Alessandro, Li Tongcang, Morandotti Roberto, Zhang Xiang, Chen Zhigang (2012) Generation of linear and nonlinear nonparaxial accelerating beams. Opt Lett 37(14):2820–2822
8. Durnin J, Miceli JJ, Eberly JH (1987) Diffraction-free beams. Phys Rev Lett 58(15):1499
9. Gori F, Guattari G, Padovani C (1987) Bessel-gauss beams. Opt Commun 64(6):491–495
10. Arlt J, Garces-Chavez V, Sibbett W, Dholakia K (2001) Optical micromanipulation using a bessel light beam. Opt Commun 197(4):239–245
11. Duocastella M, Arnold CB (2012) Bessel and annular beams for materials processing. Laser Photon Rev 6(5):607–621
12. Mitri FG (2014) Single bessel tractor-beam tweezers. Wave Motion 51(6):986–993
13. Sokolovskii GS, Dudelev VV, Losev SN, Soboleva KK, Deryagin AG, Kuchinskii VI, Sibbett W, Rafailov EU (2014) Optical trapping with bessel beams generated from semiconductor lasers. J Phys Conf Ser 572:012039. IOP Publishing
14. Krzic U, Gunther S, Saunders TE, Streichan SJ, Hufnagel L (2012) Multiview light-sheet microscope for rapid in toto imaging. Nat Methods, 9(7):730–733
15. Levoy Marc, Ng Ren, Adams Andrew, Footer Matthew, Horowitz Mark (2006) Light field microscopy. ACM Trans Graph 25(3):924–934
16. Purnapatra SB, Bera S, Mondal PP (2012) Spatial filter based bessel-like beam for improved penetration depth imaging in fluorescence microscopy. Sci Rep 2:692
17. Planchon TA, Gao L, Milkie DE, Davidson MW, Galbraith JA, Galbraith CG, Betzig E (2011) Rapid three-dimensional isotropic imaging of living cells using bessel beam plane illumination. Nat Methods, 8(5):417–423
18. Tomer R, Khairy K, Amat F, Keller PJ (2012) Quantitative high-speed imaging of entire developing embryos with simultaneous multiview light-sheet microscopy. Nat Methods 9(7):755–763
19. Prevedel R, Yoon YG, Hoffmann M, Pak N, Wetzstein G, Kato S, Schrödel T, Raskar R, Zimmer M, Boyden ES, Vaziri A, et al (2014) Simultaneous whole-animal 3D imaging of neuronal activity using light-field microscopy. Nat Methods 11(7):727–730
20. Flock ST, Jacques SL, Wilson BC, Star WM, van Gemert MJ (1992) Optical properties of intralipid: a phantom medium for light propagation studies. Lasers Surg Med, 12(5):510–519

Chapter 5
Self-suppression of Bessel Beam Side Lobes for High-Contrast Light Sheet Microscopy

An ideal illumination for light sheet fluorescence microscopy entails both a localized and a propagation invariant optical field. Both Bessel beams and Airy beams satisfy these conditions, but their non-diffracting feature comes at the cost of the presence of high-energy side lobes that notably degrade the imaging contrast and induce photobleaching. This chapter discusses the results of the article *Self-suppression of Bessel beam side lobes for high-contrast light sheet microscopy*. We demonstrate the use of a light droplet illumination whose side lobes are self-suppressed by interfering Bessel beams of specific k-vectors. Our droplet illumination (see previous Chap. 4) readily achieves a 50% extinction of the out-of-focus light, providing a more efficient energy localization and an increased imaging contrast obtained in a standard light sheet microscope. This work is currently under revision in Optica (OSA).

5.1 Introduction

One of the main challenges in optical imaging is enabling a high spatiotemporal resolution while providing an efficient light distribution across the samples. Whereas the finite numerical aperture (NA) of lenses imposes a limit on the achievable spatial resolution, the geometry that is used to illuminate and collect light from a sample balances the light distribution efficiency. Standard wide-field microscopes, for example, employ a collimated beam of light to illuminate a sample in either reflection (epi) or transmission mode. This results in a poor contrast due to the amount of out-of-focus light distributed across a sample and collected by the system. In scanning confocal microscopy, this problem has been overcome by the optical sectioning capability enabled by the use of spatial filters that match the size of the system point spread function (PSF) placed at the object conjugate plane [1]. Nevertheless, besides a relatively low temporal resolution, confocal schemes involve extended out-of-focus

G. Di Domenico, *Electro-optic Photonic Circuits*, Springer Theses,
https://doi.org/10.1007/978-3-030-23189-7_5

volumes within the sample to be fully illuminated, in turn potentially inducing sample photodamage [2].

An efficient method to enable rapid, high-contrast, volumetric imaging with minimal sample exposure is given by light sheet microscopy [3, 4]. This method involves two orthogonal objective lenses that separately perform the excitation and the detection of sample fluorescence. In particular, the sample is illuminated with a thin sheet of light and the emission is collected along the axis perpendicular to the plane of illumination, thus providing a transverse and axial resolution imposed by the the product of the illumination and collection PSFs [5–8]. Light-sheet microscopy [9–11] is particularly suited for imaging deep in transparent tissues or in whole organisms [12] at high acquisition rates and can reach a spatial resolution up to 150 nm using structured illumination [13]. Given the high collection efficiency and the improved photon balance within the samples, specimen photobleaching and phototoxicity are minimized compared to wide-field or confocal microscopy.

A major limitation in light sheet microscopy is given by the rapid divergence of the light sheets involved when focusing conventional Gaussian beams, which in turn limits the achievable field of view and leads to image artefacts across the edges. To overcome this limit, diffraction-free beams, such as Bessel beams [14] and Airy beams [15], have seen extensive use in light sheet microscopy because they exhibit a theoretically infinite depth of focus. As such, these beams can generate a non-diverging sheet of light, but this comes at the cost of a considerable amount of background light that spreads out of the object plane. Bessel beams are indeed characterized by an intensity distribution described by a Bessel function of the zero-th order, which is composed of a central peak of high intensity followed by high-energy side lobes, which significantly affect the imaging contrast and increase sample phototoxicity [8, 13].

Here we demonstrate the application in light sheet microscopy of a diffraction-free light droplet illumination [16] (Chap. 4) that manifests a self-suppression of the high energy side lobes. Light droplets are recently demonstrated non-diffracting beams of light of finite depth of focus that are obtained interfering multiple co-axial Bessel beams. Similarly to previous approaches proposed for material processing [17, 18], our method exploits interference of a set of plane waves whose k-vectors lie on two different co-axial cones. In particular, to theoretically study the phenomenon, we perform the Fourier transform of two concentric annular sources whose aspect ratio provides the maximum confinement of light along the main peak and cancel the unwanted side lobes through a selective destructive interference. Our droplet illumination is demonstrated to significantly reduce the undesired out-of-focus light contribution with respect to that given by standard Bessel and Airy beams, resulting in an enhanced imaging contrast.

5.2 Experimental Methods

Figure 5.1a shows a schematic of the optical setup. The wavefront of an expanded collimated laser beam ($\lambda = 532$ nm) was modulated by a spatial light modulator (SLM, Holoeye LC2012) that operated in amplitude mode by an appropriate rotation of the input linear polarization, which was set at 45° with respect to the $\hat{y}$ axis using the $\lambda/2$ waveplate. A linear polarizer was placed next to the SLM with the axis parallel to the y direction. The amplitude masks displayed by the SLM (Fig. 5.1b–c) consisted of concentric annular distributions of fixed outer radius ($r1$) of thickness dr and varying inner radius ($r2$). An objective lens (Olympus Plan N 10X) of NA = 0.25 was used to perform a Fourier transform of the modulated annular fields and to generate the light droplet illumination. The fluorescent signal from the sample was collected by a second objective lens (Zeiss LD A-Plan 20x Ph1 M27) of NA = 0.35 and a tube lens (Thorlabs TTL200). A bandpass filter (Thorlabs FELH0550) was used to select only the fluorescent signal, which was detected by a CCD camera (Photometrics CoolSNAP HQ2).

To find the aspect ratio r_2/r_1 that provides the maximal energy confinement, we considered the generic electric field describing a zero-order Bessel-Gauss beam [19–21]

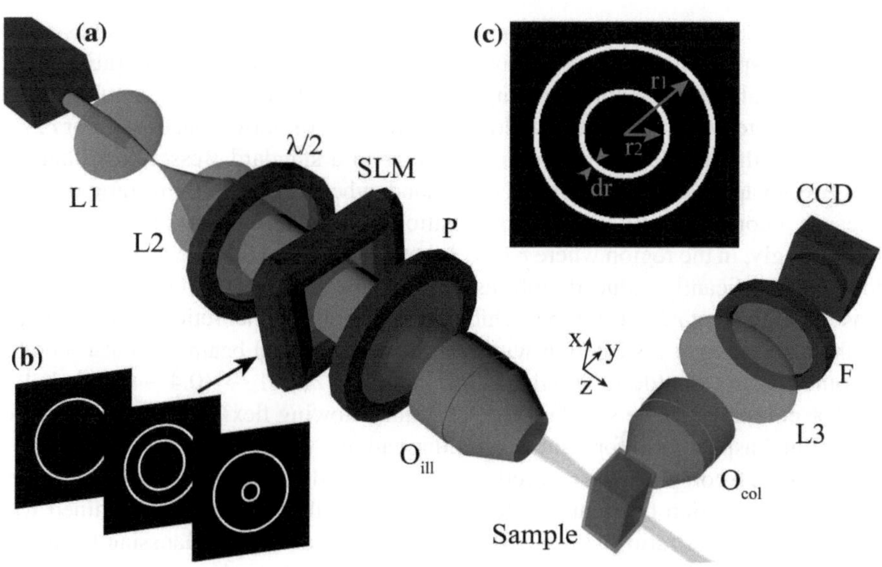

Fig. 5.1 **a** Schematic of the experimental setup. An expanded laser beam was modulated by an SLM and focused by an objective lens (O_{ill}) of NA = 0.25. Fluorescence light was collected by a second orthogonal objective (O_{col}) of NA = 0.35 and a tube lens L3, and detected by a CCD camera. L_i lenses; $\lambda/2$ half-wave plate; P linear polarizer; F bandpass filter. **b–c** A set of amplitude binary masks composed of two concentric rings of inner and outer radii, $r2$ and $r1$ respectively, and thickness dr displayed by the SLM

$$E(r', z, \theta) = E_0 \frac{w_0}{w(z)} \exp\left[i\left(\left(k - \frac{k_r(\theta)^2}{2k}\right)z - \Phi(z)\right)\right]$$
$$\times J_0\left[\frac{r' k_r(\theta)}{\frac{iz}{z_R}+1}\right] \exp\left[\left(\frac{ik}{2R(z)} - \frac{1}{w^2(z)}\right)\left(r'^2 + \frac{k_r(\theta)^2 z^2}{k^2}\right)\right] \tag{5.1}$$

where J_0 is a Bessel function of the first kind and zero order, $r' = \sqrt{x^2+y^2}$ is the radial coordinate, $k_r(\theta) = k\sin(\theta)$ and $k_z(\theta) = k\cos(\theta)$ are the radial and longitudinal wavevectors respectively (being $k = |\mathbf{k}| = \sqrt{k_r(\theta)^2 + k_z(\theta)^2} = 2\pi/\lambda$ and $\theta = \text{atan}(r/f)$ the angle between the wavevector and the propagation axis), $z_r = \pi n w_0^2/\lambda$ is the Rayleigh length, w_0 is a measure of the width of the Gaussian term, $w(z) = w_0\sqrt{1+(z/z_r)^2}$, $R(z) = z/(1+z_r^2)$ is the radius of curvature of the beam's wavefronts and $\Phi(z) = \tan^{-1}(z/z_r)$ is the Gouy phase shift. Thus, the total optical power distributed along the droplet side lobes for given values of θ_1 and θ_2 is given by

$$P = \int_{r'_{min}}^{\infty} 2\pi r' |E_1(r', z, \theta_1) + E_2(r', z, \theta_2)|^2 dr', \tag{5.2}$$

where r'_{min} is the radial distance associated to the first minimum of the intensity pattern.

Figure 5.2a shows a plot of the theoretical and measured droplet side lobes power (P) with respect to the Bessel beam power (P_B) as a function of the ratio $r2/r1$. Experimental data was obtained varying the radius $r2$ of the inner annulus on the SLM and detecting the resulting field distribution after Fourier transform. For $r2/r1$ close to 1, the droplet beam profile is similar to a standard Bessel beam and the power ratio approaches to 1. On the other hand, when $r2/r1 \to 0$ this ratio diverges because it corresponds to the superimposition of a Bessel beam with a plane wave. Interestingly, in the region where $r2/r1$ is between 0.4 and 0.8, the power in the side lobes is significantly reduced with respect to that of a Bessel beam. The minimum was found to be at $r2/r1 = 0.57$, which corresponds to a theoretical 62% reduction of the off-axis light distribution with respect to the Bessel beam. It is also notable that the amount of side lobe light within the range $r2/r1 \to (0.4 - 0.8)$ is below $\sim$ 50% compared to the standard Bessel beam, allowing flexibility in the choice of the optimal aspect ratio for light sheet illumination.

Figure 5.2b compares the theoretical 2D intensity distribution (xz plane) given by a standard Gaussian beam and a Bessel beam with the droplet field obtained with the optimal aspect ratio ($r2/r1 = 0.57$). Unlike the diverging Gaussian beam, the Bessel beam manifests an invariant transverse profile along the propagation axis $\hat{z}$. Nevertheless, half of its total energy is distributed along the side lobes. By contrast, our droplet illumination maintains an extended depth of focus, ~2.6 times greater than the one given by the Gaussian beam, but provides a significantly enhanced energy confinement across the central peak without loss in transverse resolution.

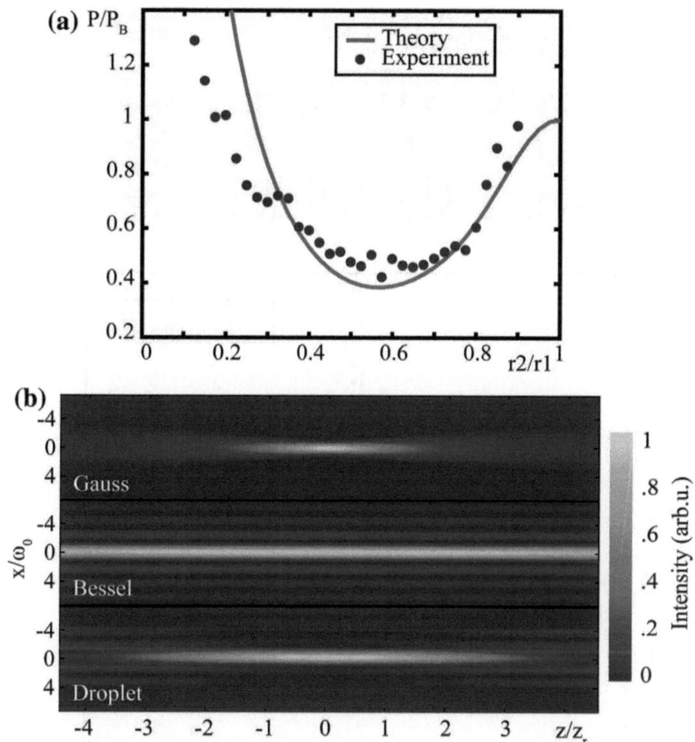

Fig. 5.2 **a** Plot of the ratio between optical power of the droplet side lobes with respect to the Bessel lobe power as a function of the ratio r_2/r_1. Solid line is the theoretical prediction while red dots are experimental data. **b** Top-view (xz plane) intensity profiles of a Gaussian, a Bessel and a Droplet beam as a function of the number of Rayleigh lengths

Figure 5.3a shows the measured intensity profiles for both a standard Bessel (top) and a droplet beam (bottom). While the Bessel beam was obtained by Fourier transform of a single annular aperture ($r1 = 1.7$ mm diameter and 150 μm width) displayed on the SLM, the illustrated droplet beam resulted from the Fourier transform of two annuli in the case of $r2/r1 = 0.57$, where $r1$ was maintained equal and the amplitude normalized to that of the reference Bessel beam. The associated normalized radial intensity profiles for both beams are illustrated in Fig. 5.3b. Both profiles show a central peak of comparable width. Nevertheless, the droplet beam manifests significantly lower side lobes as a result of the selective destructive interference of two fields of slightly different k-vectors. Integrating the intensity along the radial distribution of the normalized Bessel and droplet beams, we found an overall 57.9% reduction in the side lobes intensity, whilst the intensity of the first side lobes decreased by 88%.

The enhanced imaging capability offered by the droplet illumination was demonstrated in a standard light sheet microscope on a fluorescent specimen (Fig. 5.1a).

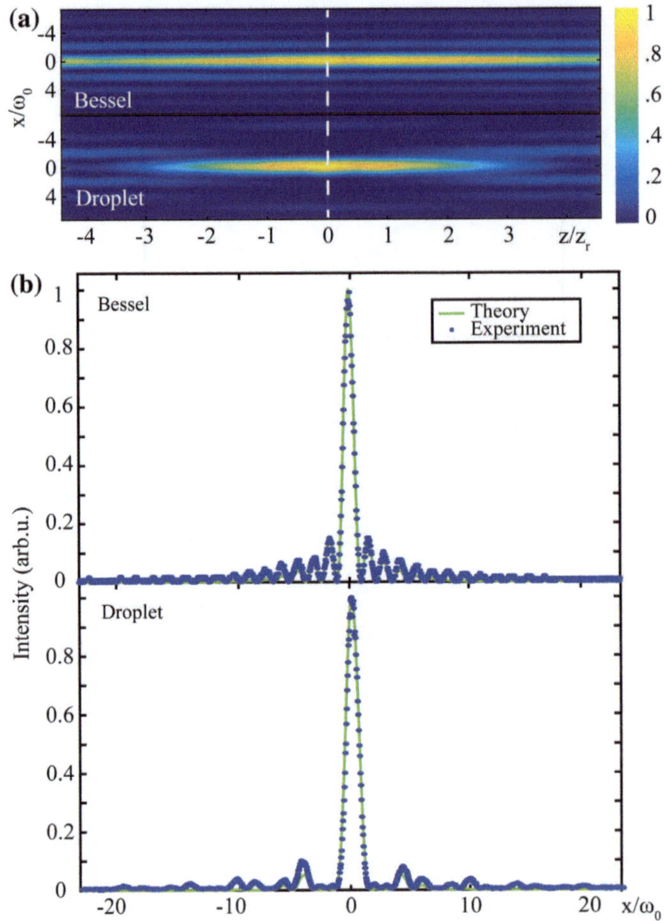

Fig. 5.3 **a** Representative Bessel beam obtained with a single ring of radius r_1 (top) and droplet beam with ratios $r2/r1 = 0.57$ (bottom). **b** Measured radial intensity profile of the beams along the dashed line together with the theoretical prediction (green line)

This was composed of fluorescent spheres with diameters ranging between 5 and 50 μm embedded in a fixed gel mounted on a glass cuvette. Figure 5.4 shows the images (in a log scale) obtained with a standard Bessel beam (Fig. 5.4a) and the droplet illumination (Fig. 5.4b). The images were acquired by scanning the sample along the $\hat{y}$ direction and integrating the emitted fluorescent signal acquired by the CCD camera. The image obtained using a conventional Bessel beam appears with a significant background across the field of view as a consequence of the out-of-focus-light given by the side lobes that is collected by the system. In turn, the associated image obtained using the droplet illumination has a significantly reduced background, the result of the side lobes self-suppression capability of these beams.

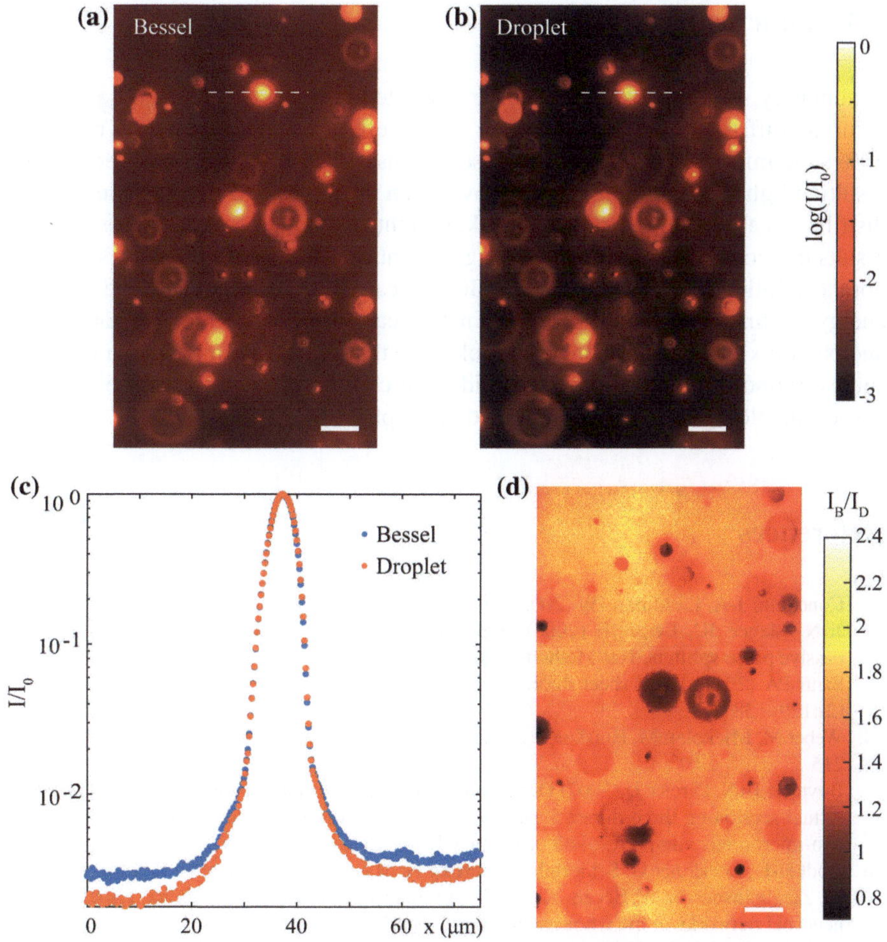

Fig. 5.4 Images **a** and **b** were obtained by a light sheet microscope using a Bessel beam and a droplet beam illumination respectively. **c** Intensity profiles along the dashed lines from **a** and **b**. Scale bar is 20 μm. Image **d** maps the intensity ratio between the images obtained with Bessel and droplet illumination

Figure 5.4c illustrates a line profile along a sphere captured across the image field of view, which emphasize the enhanced contrast given by the droplet illumination. To give a measure of the background reduction, Fig. 5.4d illustrates the ratio between the frames obtained with the Bessel and the droplet illuminations. This ratio is close to 1 near the fluorescent spheres and grows elsewhere, indicating an averaged two-fold reduction of the background illumination.

5.3 Conclusion

In summary, we demonstrated that the unwanted high-energy side lobes present in common diffractionless optical beams can be effectively suppressed by more than 50% superimposing two co-axial Bessel beams of slightly different k-vectors. The resulting light droplet illumination provides an improved and more efficient energy distribution along the central main peak. In light sheet fluorescence microscopy, this results in a considerably increased imaging contrast without affecting the spatiotemporal resolution offered by a standard Bessel beam illumination. Given the improved energy confinement, the droplet illumination can further limit both photobleaching and phototoxicity of the fluorescent molecules that induce a sample degradation. As such, our findings pave the way to rapid, high-contrast volume imaging of cells and tissues, heralding novel in-vivo biomedical applications.

References

1. Conchello J-A, Lichtman JW (2005) Optical sectioning microscopy. Nat Methods 2(12):920
2. Ji N, Magee JC, Betzig E (2008) High-speed, low-photodamage nonlinear imaging using passive pulse splitters. Nat Methods 5(2):197–202
3. Santi PA (2011) Light sheet fluorescence microscopy: a review. J Histochem Cytochem 59 (2):129–138
4. Weber M, Mickoleit M, Huisken J (2014) Light sheet microscopy. Methods Cell Biol 123:193–215
5. Buytaert JAN, Dirckx JJ (2007) Design and quantitative resolution measurements of an optical virtual sectioning three-dimensional imaging technique for biomedical specimens, featuring two-micrometer slicing resolution. J Biomed Opt 12(1):014039
6. Dodt H-U, Leischner U, Schierloh A, Jährling N, Mauch CP, Deininger K, Deussing JM, Eder M, Zieglgänsberger W, Becker K (2007) Ultramicroscopy: three-dimensional visualization of neuronal networks in the whole mouse brain. Nat Methods 4(4):331
7. Keller PJ, Dodt H-U (2012) Light sheet microscopy of living or cleared specimens. Curr Opin Neurobiol 22(1):138–143
8. Keller PJ, Schmidt AD, Wittbrodt J, Stelzer EHK (2008) Reconstruction of zebrafish early embryonic development by scanned light sheet microscopy. Science 322(5904):1065–1069
9. Breuninger T, Greger K, Stelzer EHK (2007) Lateral modulation boosts image quality in single plane illumination fluorescence microscopy. Opt Lett 32(13):1938–1940
10. Keller PJ, Schmidt AD, Santella A, Khairy K, Bao Z, Wittbrodt J, Stelzer EHK (2010) Fast, high-contrast imaging of animal development with scanned light sheet-based structured-illumination microscopy. Nat Methods 7(8):637–642
11. Neil MAA, Juškaitis R, Wilson T (1997) Method of obtaining optical sectioning by using structured light in a conventional microscope. Opt Lett 22(24):1905–1907
12. Ahrens MB, Orger MB, Robson DN, Li JM, Keller PJ (2013) Whole-brain functional imaging at cellular resolution using light-sheet microscopy. Nat Methods 10(5):413–420
13. Chen B-C, Legant WR, Wang K, Shao L, Milkie DE, Davidson MW, Janetopoulos C, Wu XS, Hammer JA, Liu Z et al. (2014) Lattice light-sheet microscopy: imaging molecules to embryos at high spatiotemporal resolution. Science 346(6208):1257998
14. Planchon TA, Gao L, Milkie DE, Davidson MW, Galbraith JA, Galbraith CG, Betzig E (2011) Rapid three-dimensional isotropic imaging of living cells using Bessel beam plane illumination. Nat Methods 8(5):417–423

15. Vettenburg T, Dalgarno HIC, Nylk J, Coll-Lladó C, Ferrier DEK, Čižmár T, Gunn-Moore FJ, Dholakia K (2014) Light-sheet microscopy using an Airy beam. Nat Methods 11(5):541–544
16. Antonacci G, Di Domenico G, Silvestri S, DelRe E, Ruocco G (2017) Diffraction-free light droplets for axially-resolved volume imaging. Sci Rep 7(1):17
17. He F, Yu J, Tan Y, Chu W, Zhou C, Cheng Y, Sugioka K (2017) Tailoring femtosecond 1.5-μm bessel beams for manufacturing high-aspect-ratio through-silicon vias. Sci Rep 7:40785
18. Mori S (2015) Side lobe suppression of a Bessel beam for high aspect ratio laser processing. Precis Eng 39:79–85
19. Durnin JJ, Miceli Jr JJ, Eberly JH (1987) Diffraction-free beams. Phys Rev Lett 58(15):1499
20. Gori F, Guattari G, Padovani C (1987) Bessel-gauss beams. Opt Commun 64(6):491–495
21. Sheppard CJR, Wilson T (1978) Gaussian-beam theory of lenses with annular aperture. IEE J Microw Opt Acoust 2(4):105–112

Chapter 6
Microscopic Reversibility, Nonlinearity, and the Conditional Nature of Single Particle Entanglement

In 1935 Erwin Schrödinger [1] motivated by the famous paper [2] from Einstein, Podolsky and Rosen (EPR) comment that *"Maximal knowledge of a total system does not necessarily include total knowledge of all its parts, not even when these are fully separated from each other and at the moment are not influencing each other at all."* and he coined the term "entanglement of our knowledge" to describe the inconsistency between the local realism and a quantum mechanical explanation of realty, what Einstein instead terms "quantum-nonlocality". In 1964 John Bell [3] quantitatively express this violation of local realism obtaining certain bounds (Bell inequalities) for the correlations of measurements on a two-particle systems, that can be directly measured. During the past few decades, there has been attempt to find circumstances in which a single particle is able to show quantum non-locality.

In this chapter we discuss the problems connected to the idea of entangling two distant system by a single particle in various quantitative ways. We found the goal to be incompatible with microscopic reversibility principle and we also show through detailed calculation of the final state, how the two distant system are not entangled in such a way as to be useful for quantum information technology (see Appendix A). We also show how entanglement requires a simultaneous local interaction between all input and output particles, so that, on consequence of microscopic reversibility, entanglement is associated to the spontaneous increase in the number of particles in an isolated system. Our result unveils a fundamental limit in schemes suggesting one particle entanglement, confirms that there is no means for a single particle to entangle two distant systems with anything close to certainty, even in the most ideal of cases and clarifies the long standing debate on the limits of using single photons and linear optics in implementing quantum technology.

The work is, at the present time, unpublished and it is currently under revision in Physical Review Letters.

G. Di Domenico, *Electro-optic Photonic Circuits*, Springer Theses,
https://doi.org/10.1007/978-3-030-23189-7_6

6.1 Introduction

Quantum mechanics implies physical systems that manifest nonlocality, i.e., that violate local realism [4–8]. Such systems must by definition support at least two interactions with macroscopic measuring devices at a distance. The set of predictable results for the joint measurements of two observables $\mathcal{X}_1$ and $\mathcal{X}_2$, based on local hidden-variable theories, translates into appropriate Bell-inequalities that can be experimentally tested [9]. When these are violated, the state is said to be entangled, because in terms of the eigenstates of the two distant measuring apparatus the wavefunction cannot be factored out, i.e., $|\psi\rangle \neq |\psi_1(\mathcal{X}_1)\rangle|\psi_2(\mathcal{X}_2)\rangle$. As a consequence, since macroscopic phenomena obey local realism, entangled states allow an extension of technology into what is termed quantum technology, simply because they manifest effects that are impossible in the macroscopic domain. Examples of quantum technology are dense coding, teleportation, and quantum computing [10]. A fundamental issue that has a key practical consequence consists in the possibility of generating entanglement by means of a single particle, for example, a single photon split by a beam-splitter. If this were possible, the important result would be that quantum technology could be designed without the involvement of rare multi-particle interaction processes, i.e., in optics, without non-linear optics for ultra-weak beams [11, 12].

Understandably, the issue has a long trail of arguments that dates back to the Bohr–Einstein debate.[1] Einstein was unable to put nonlocality on physical footing using a single-photon gedanken experiment, a difficulty later corroborated by the apparent validity of single-particle hidden-variable theories [13]. The nonlocal features of quantum theory were introduced using a second auxiliary particle, so that nonlocality arises as a property of a combination of two particles that suffered a previous interaction (the Einstein–Podolski–Rosen—EPR effect) [2]. Examples are the decay of a positronium atom into two photons [14–16] and the spontaneous down-conversion of an UV photon into two infrared photons in a crystal [17, 18]. A basic step forward, compared to the original Bohr–Einstein debate, is represented by attempts to investigate single-particle nonlocality and entanglement using photon-homodyne-detection [19–25]. The key to these attempts, compared to the original Bohr–Einstein debate, is the use of homodyne detection [19, 22, 24], as suggested by [21, 25]. Unfortunately, as discussed by Greenberger et al. in [26], all schemes using homodyne detection are based on the measurement of phase through interference with a reference oscillator, so that the original photon and the reference photons are indistinguishable, the number of photons is itself indeterminate, and hence entanglement based on these schemes does not demonstrate nonlocality for a single photon. Moreover, the use of a single phase-reference as local oscillator on both distant systems means that these have a level of superposition, so that the two local states are not in fact describable as different parts [13]. The question therefore lies open if two distant systems can be entangled by a single particle.

[1] A complete description of the conceptual development of the notion of nonlocality can be found in [8].

6.1.1 Particle State

We address single-particle splitting using the separability criterion, the most widely accepted definition of entanglement for bipartite states [7]: the state of a system composed by two separate subsystems A and B is entangled only if it is not separable, i.e., only if its density matrix cannot be expressed in the form

$$\rho_{AB} = \sum_i p_i \rho_A^i \otimes \rho_B^i, \tag{6.1}$$

where the p_i's are probability coefficients, and the ρ_A^i's and ρ_B^i's are density matrices referring to the two subsystems A and B. This definition is general, in the sense that it holds both for mixed states and for pure states, in which case one of the p_i's is 1, all the others vanish, and the surviving ρ_A and ρ_B refer to pure states of A and B, i.e.,

$$\rho_{AB} = |\psi_A\rangle\langle\psi_A| \otimes |\psi_B\rangle\langle\psi_B|. \tag{6.2}$$

This separability criterion is directly linked to the development of quantum information technology. This is because the key ingredients that allow non-classical tasks are quantum phase factors, and these are completely absent in the case represented by Eq. (6.1) [7]. Put differently, it is the violation of Eq. (6.1) that allows us to construct the exclusively quantum superposition of states that grows more than linearly with the number of subsystems A, B, C .... .

In physical terms, a density matrix of the kind given in Eq. (6.1) is consistent with the existence of a *hidden variable* characterized by the probability coefficients p_i, so that no Bell's inequality violation is present: the information content of any quantum state represented by Eq. (6.1) is completely classical, so that no quantum algorithms can be implemented. As well known the converse is false, in that not all states not satisfying Eq. (6.1) can be exploited for quantum technology [27].

6.2 Thought Experiment

Coming now back to our system, the paradigm for the supposed entanglement of two distant bodies is a single photon entangling two distant two-level atoms A and B. The reasoning runs as follows (see, for example, [28]): Let a photon impinge on a beam-splitter (BS), so that its state $|\phi\rangle$ after the BS can be written as

$$|\phi\rangle = \frac{1}{\sqrt{2}}\left[|1_1\rangle|0_2\rangle + |0_1\rangle|1_2\rangle\right], \tag{6.3}$$

or, equivalently,

$$|\phi\rangle = \frac{1}{\sqrt{2}}\left[|1\rangle + i|2\rangle\right]. \tag{6.4}$$

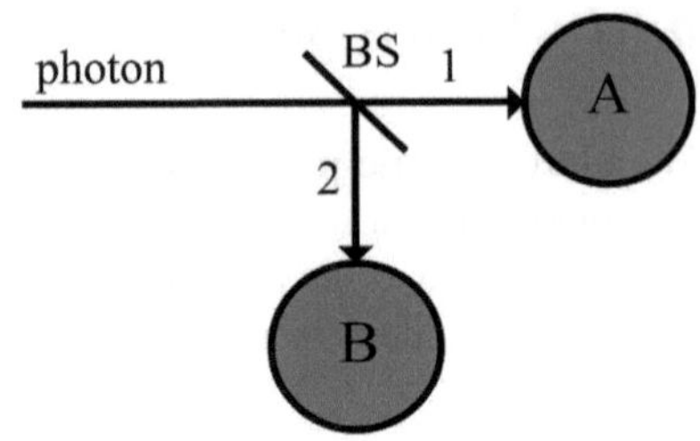

Fig. 6.1 General set-up: a photon split into two modes 1 and 2 supposedly entangling two distant atoms A and B

In Eq. (6.3) the subscripts 1 and 2 label forward and downward modes and the digits in the kets are the corresponding occupation numbers, while, in Eq. (6.4), $|1\rangle$ and $|2\rangle$ are forward and downward modes (Fig. 6.1). The two modes are not entangled "per se", as made clear by the two following statements [28]:

(I) While Eq. (6.3) appears to represent $|\phi\rangle$ as a nonproduct state, this representation can be reduced to Eq. (6.4), i.e., to the superposition of two eigenstates, which is not entangled;
(II) The presence of a single photon intrinsically rules out the possibility of carrying out two (or more) measurements at space-like distances, this being the basic requirement for any observation of nonlocality or Bell-inequality violation.

To overcome these obstacles, the two modes can be coupled with ideal efficiency to two distant equal two-level subsystems A (atom 1) and B (atom 2) (Fig. 6.1) with resonant ground $|g\rangle$ and excited $|e\rangle$ levels, both initially in $|g\rangle$. The final state is apparently given by

$$|\psi\rangle = \frac{1}{\sqrt{2}} \left[|e\rangle_1 |g\rangle_2 + |g\rangle_1 |e\rangle_2 \right], \tag{6.5}$$

which cannot be reduced into any other representation [28] and appears to be entangled. The issue has been discussed by several authors (see, for example, [29, 30]). In particular, in [29] the author casts some doubt on the conclusions of [28], that is, that the state of Eq. (6.5) is entangled, claiming that, if we project Eq. (6.5) on the position state of one of the two atoms, the state of the system A, B is not entangled. In [31], Van Enk replies that no projection operation, i.e., no position measurement, is present in his scheme.

We here observe that Eq. (6.5) is somewhat misleading. In fact, a correct description of a quantum system requires to explicitly consider a complete set of observables. In other words, the complete representation must obey microscopic reversibility, which amounts, in our case, to the conservation of total energy, momentum, and angular momentum of our system. In this frame, we note that the role of atom recoil in limiting entanglement fidelity has been previously considered in the conceptually reversed situation in which light is emitted from distant atoms (see, e.g., [32, 33]).

To provide a complete representation of the state of A and B, we suppose, without loss of generality, that a photon circularly polarized in the forward z-direction impinges on a hydrogen atom 1. If excitation takes place, the atomic state is correctly expressed as $|2, 1, +1, \varphi_{1,e}\rangle$, where 2, 1, +1 respectively refers to the ordinary quantum numbers n, l, m and $\varphi_{1,e}$ is the wavefunction of the center of mass of atom 1, after absorbing the photon momentum. If atom 1 does not interact with the photon, its state is given by $|1, 0, 0, \varphi_{1,g}\rangle$, where $\varphi_{1,g}$ is the wavefunction of the center of mass of atom 1 not interacting with the photon. The same holds for atom 2. Thus, the complete representation of the atomic state is not given by Eq. (6.5), but reads

$$|\Psi\rangle = \frac{1}{\sqrt{2}}[|2, 1, +1, \varphi_{1,e}\rangle|1, 0, 0, \varphi_{2,g}\rangle + |1, 0, 0, \varphi_{1,g}\rangle|2, 1, +1, \varphi_{2,e}\rangle]. \quad (6.6)$$

In turn, Eq. (6.6) implies that the average value of any observable $\widehat{O}$ reads

$$\begin{aligned}\langle\Psi|\widehat{O}|\Psi\rangle = \frac{1}{2} \\ \{\left[\langle 2, 1, +1, \varphi_{1,e}|\langle 1, 0, 0, \varphi_{2,g}|\right] \widehat{O} \left[|2, 1, +1, \varphi_{1,e}\rangle|1, 0, 0, \varphi_{2,g}\rangle\right] + \\ \left[\langle 2, 1, +1, \varphi_{1,e}|\langle 1, 0, 0, \varphi_{2,g}|\right] \widehat{O} \left[|1, 0, 0, \varphi_{1,g}\rangle|2, 1, +1, \varphi_{2,e}\rangle\right] + \\ \left[\langle 1, 0, 0, \varphi_{1,g}|\langle 2, 1, +1, \varphi_{2,e}|\right] \widehat{O} \left[|2, 1, +1, \varphi_{1,e}\rangle|1, 0, 0, \varphi_{2,g}\rangle\right] + \\ \left[\langle 1, 0, 0, \varphi_{1,g}|\langle 2, 1, +1, \varphi_{2,e}|\right] \widehat{O} \left[|1, 0, 0, \varphi_{1,g}\rangle|2, 1, +1, \varphi_{2,e}\rangle\right]\}. \quad (6.7)\end{aligned}$$

Consider now any experiment aimed at revealing a supposed Bell-inequality violation dealing with observables $\widehat{O}_i$ related to angular momenta. In this case the second and third terms in Eq. (6.7) vanish, because bras and kets with different internal energies remain orthogonal when $\widehat{O}_i$ is applied. Therefore, we have

$$\begin{aligned}\langle\Psi|\widehat{O}_i|\Psi\rangle = \frac{1}{2}\{\left[\langle 2, 1, +1, \varphi_{1,e}|\langle 1, 0, 0, \varphi_{2,g}|\right] \widehat{O}_i \left[|2, 1, +1, \varphi_{1,e}\rangle|1, 0, 0, \varphi_{2,g}\rangle\right] + \\ \left[\langle 1, 0, 0, \varphi_{1,g}|\langle 2, 1, +1, \varphi_{2,e}|\right] \widehat{O}_i \left[|1, 0, 0, \varphi_{1,g}\rangle|2, 1, +1, \varphi_{2,e}\rangle\right]\}. \\ (6.8)\end{aligned}$$

As a consequence,

$$\langle\Psi|\widehat{O}_i|\Psi\rangle = Tr\left[\widehat{\rho}\,\widehat{O}_i\right], \quad (6.9)$$

with

$$\begin{aligned}\widehat{\rho} = \frac{1}{2}\{\left[|2, 1, +1, \varphi_{1,e}\rangle|1, 0, 0, \varphi_{2,g}\rangle\right]\left[\langle 2, 1, +1, \varphi_{1,e}|\langle 1, 0, 0, \varphi_{2,g}|\right] + \\ \left[|1, 0, 0, \varphi_{1,g}\rangle|2, 1, +1, \varphi_{2,e}\rangle\right]\left[\langle 1, 0, 0, \varphi_{1,g}|\langle 2, 1, +1, \varphi_{2,e}|\right]\}. \quad (6.10)\end{aligned}$$

The density matrix given by Eq. (6.10) corresponds to a separable *mixed* state of the kind of Eq. (6.1) with

$$\begin{aligned}p_1 = p_2 = \frac{1}{2}, \\ \rho_A^1 = |2, 1, +1, \varphi_{1,e}\rangle\langle 2, 1, +1, \varphi_{1,e}|,\end{aligned}$$

$$
\begin{aligned}
\rho_B^1 &= |1, 0, 0, \varphi_{2,g}\rangle\langle 1, 0, 0, \varphi_{2,g}|, \\
\rho_A^2 &= |1, 0, 0, \varphi_{1,g}\rangle\langle 1, 0, 0, \varphi_{1,g}|, \\
\rho_B^2 &= |2, 1, +1, \varphi_{2,e}\rangle\langle 2, 1, +1, \varphi_{2,e}|.
\end{aligned} \tag{6.11}
$$

More precisely, all experiments that involve correlations between the internal atomic degrees of freedom of the atoms are described by tracing over the internal energy, leading to the disentangled mixed state of Eq. (6.10).

6.3 Discussion

Our result agrees with the statement that there are only two ways to entangle A and B:

(I) We let A and B interact with one another and then we separate them [7, 10].
(II) We transfer the entanglement previously present in another two-particle system on the already remote A and B (see, e.g., [32, 33]). In this case, no tracing process leading to disentanglement takes place.

To enforce the latter statement, we refer to a parametric-fluorescence process by which two photons a, b, entangled through their total vanishing spin and with common angular frequency ω, are emitted in opposite directions. The two photons a, b are respectively coupled with ideal efficiency to two equal two-level atoms A (atom 1) and B (atom 2) resonant at ω, so that the final (entangled) state $|\Psi_f\rangle$ reads

$$
|\Psi_f\rangle = \frac{1}{\sqrt{2}}[|2, 1, +1, \varphi_{1,e}\rangle|2, 1, -1, \varphi_{2,e}\rangle + |2, 1, -1, \varphi_{1,e}\rangle|2, 1, +1, \varphi_{2,e}\rangle]. \tag{6.12}
$$

The main difference between the state $|\Psi\rangle$ of Eq. (6.6) and the state $|\Psi_f\rangle$ of Eq. (6.12) is that the latter cannot be disentangled by tracing over the internal energy. In other words, referring to quantum algorithms, a split photon does not entangle two distant atoms A, B, while two initially entangled photons can integrally transfer their entanglement to A and B.

We underline how Eq. (6.10) does not rely on any *ad hoc* assumption, but directly follows from first principles of quantum mechanics: the absence of entanglement between the internal degrees of freedom of two distant and initially uncorrelated atoms persists, after interaction with a single split photon, as a consequence of the mere existence of two different internal energy levels, even if these are *not observed.* Furthermore, our result is not specific to photons and atoms. It holds equally well for other single particles and pairs of distant microscopic bodies. For example, an electron split along two paths using an appropriate Stern-Gerlach-type experiment and sent to two distant bodies will not entangle their internal degrees of freedom.

We note that useful entanglement is only remotely connected to the mere existence of correlations. For example, if we know classically that only one object is either sent to output A or B, its detection at A tells us that it is not present at B, and viceversa, but this correlation has simply nothing to do with entanglement. However, the criterion based on entanglement potential is not directly related to quantum technology applications. In fact, as we have shown, although a single photon possesses a finite degree of entanglement potential, it cannot be exploited for quantum technology even after letting it interact with two distant atoms after being split.

Interestingly, the fact that single photons cannot entangle distinct microscopic bodies also means that they cannot enact useful quantum computing. To see this, consider a basic two-port input two-port output gate, where input and output ports are labelled 1i, 2i and 1o, 2o respectively. Assume that $|0\rangle_{1i}|0\rangle_{2i} \rightarrow |0\rangle_{1o}|f(0)\rangle_{2o}$, and $|1\rangle_{1i}|0\rangle_{2i} \rightarrow |1\rangle_{1o}|f(1)\rangle_{2o}$ where f is the desired function carried out on the target system 2 on the basis of the data input 1, and the data qubit is preserved to allow reversibility. It must be that $(\alpha|0\rangle_{1i} + \beta|1\rangle_{1i})|0\rangle_{2i} \rightarrow (\alpha|0\rangle_{1o}|f(0)\rangle_{2o} + \beta|1\rangle_{1o}|f(1)\rangle_{2o})$ and not $(\alpha|0\rangle_{1i} + \beta|1\rangle_{1i})|0\rangle_{2i} \rightarrow (\alpha|0\rangle_{1o} + \beta|1\rangle_{1o})(\alpha|f(0)\rangle_{2o} + \beta|f(1)\rangle_{2o})$, as this latter outcome would allow the cloning of an unknown input state [34]. Thus, since the ports 1 and 2 are distinguishable, the outcome of a quantum gate manifests entanglement, a property that cannot be supported by a single photon. It follows that quantum computing based on photons must necessarily involve rare multi-particle interaction processes, such as nonlinear optical phenomena at ultra-weak intensities [11, 12].

Analogously, single photons cannot be used to achieve total teleportation, where entanglement is also necessary, using linear processes [35, 36]. In agreement with our result, in [35] the authors obtain total teleportation of a known state [37], and in [36], where teleportation truly refers to an unknown state, it is obtained with a success rate of 50%. As for quantum computing, also total teleportation of an unknown state requires nonlinear schemes [38, 39]. Therefore, both quantum computing and teleportation corroborate our finding: *the information content of a single particle, whatever its quantum state, is completely classical in nature.* We underline that the validity of this statement is limited to particles involved in linear processes. In fact, a photon of frequency ω can excite two distant two-level atoms resonating at $\omega/2$, thus entangling them through a nonlinear process [40].

We note that our analysis refers to the state of a final system composed of two distant initially uncorrelated atoms after they have interacted with a single split photon. Although we have demonstrated that this final state is separable, and hence not useful for quantum algorithms, the original split photon in fact possesses a finite degree of entanglement potential, a quantification of nonseparability that only holds for single-mode photons based on the amount of correlations present between the two outputs A and B arising from the beam-splitter [41].

In this context, there is an alternative definition of entanglement relative to photons for a single-mode state σ (entanglement potential) based on the amount of correlations present between the two outputs A and B arising from a beam-splitter invested by σ [41]. Accordingly, it can be shown that only coherent states and their mixtures do not have entanglement potential. This is appealing, since coherent states are selected as completely classical; obviously classical fields do not exhibit correlations between A and B. The two criteria are equivalent, in that a single-mode photonic state σ entangled according to one of the two criterion is as well entangled for the other one: more precisely, entanglement potential can be viewed as a quantitative measurement of the amount of separability absence. However, separability is more general, since it can be applied to bipartite systems not related to photons.

6.4 Conclusion

Our discussion on entanglement and the number of particles clears the way to a more profound understanding of the phenomenon. We see that the key is not a single particle nonlocality, nor is it some hidden property of two or more particles. It is the consequence of the microscopic effect of a nonlinearity that changes the original number of particles.

So the basic question we should focus on is not the number of particles involved, but what changes as the number of particles changes or in a more general way, the number of (distinguishable) particles indicates the number of different space-like experiments that can be carried out. The result is that the number of particles flags an extended susceptibility to independent experiments. However, since our knowledge of the system is maximally described by the original wavefunction, although more particles make more independent experiments possible, their possible outcomes must not provide extra information, and hence the results must be correlated. In turn the wavefunction, which must be written so as to describe the attainable results with the given extended experimental apparatus, in terms of the new Fock-space, must be entangled. However, when the number of particles does not increase, there is no possibility of carrying out independent measurements at space-like distances, and the wavefunction need not take the form of excluding possible outcomes, these not being in any case possible. In other words, for a single photon, the increase of Fock-space or of available paths does not increase the possibility of attaining information on the photon.

To conclude *entanglement is a product of the principles of quantum mechanics applied to a system whose number of particles changes*. In its simplest form, entanglement has to do with the nonlinear process by which a single particle leads to two distinct particles.

Total entanglement can only be achieved at the expense of having the two systems interact simultaneously and directly with the entangling particle, so they cannot be already distant. The requirement of direct interaction between the two systems and the particle is corroborated by the constraints of microscopic reversibility, leading to the conclusion that what is entangled is an isolated system whose number of distinguishable particles spontaneously changes Appendix A.1.3. In this frame, the origin of non-locality follows from the microscopic reversibility applied to systems whose number of particles susceptible to space-like measurements increases.

References

1. Schrödinger Erwin (1935) The present situation in quantum mechanics. Sci Nat 23(68):807–812
2. Einstein Albert, Podolsky Boris, Rosen Nathan (1935) Can quantum-mechanical description of physical reality be considered complete? Phys Rev 47(10):777
3. John S (1964) Bell. On the Einstein Podolsky Rosen paradox
4. Bennett CH, DiVincenzo DP (2000) Quantum information and computation. Nature, 404(6775):247
5. Brunner Nicolas, Cavalcanti Daniel, Pironio Stefano, Scarani Valerio, Wehner Stephanie (2014) Bell nonlocality. Rev Mod Phys 86(2):419
6. Eisert J (2006) Entanglement in quantum information theory. PhD thesis, University of Potsdam
7. Horodecki Ryszard, Horodecki Paweł, Horodecki Michał, Horodecki Karol (2009) Quantum entanglement. Rev Mod Phys 81(2):865
8. Jammer M (1974) Philosophy of quantum mechanics. The interpretations of quantum mechanics in historical perspective. Wiley-Interscience Publications
9. Aspect Alain, Grangier Philippe, Roger Gérard (1982) Experimental realization of Einstein-Podolsky-Rosen-Bohm gedankenexperiment: a new violation of bell's inequalities. Phys Rev Lett 49(2):91
10. Preskill J (1988) Lecture notes on quantum computation. http://www.theory.caltech.edu/~preskill/ph219/index.html#lecture
11. Reiserer Andreas, Kalb Norbert, Rempe Gerhard, Ritter Stephan (2014) A quantum gate between a flying optical photon and a single trapped atom. Nature 508(7495):237–240
12. Tiecke TG, Thompson JD, de Leon NP, Liu LR, Vuletić V, Lukin MD (2014) Nanophotonic quantum phase switch with a single atom. Nature 508(7495):241–244
13. Vaidman Lev (1995) Nonlocality of a single photon revisited again. Phys Rev Lett 75(10):2063
14. Crater HW, Wong CY, Van Alstine P (2006) Tests of two-body Dirac equation wave functions in the decays of quarkonium and positronium into two photons. Phys Rev D 74(5):054028
15. Deutsch Martin (1951) Evidence for the formation of positronium in gases. Phys Rev 82(3):455
16. Deutsch Martin, Dulit Everett (1951) Short range interaction of electrons and fine structure of positronium. Phys Rev 84(3):601
17. Kwiat PG, Mattle K, Weinfurter H, Zeilinger A, Sergienko AV, Shih Y (1995) New high-intensity source of polarization-entangled photon pairs. Phys Rev Lett 75(24):4337
18. Shih YH, Sergienko AV, Rubin MH, Kiess TE, Alley CO (1994) Two-photon entanglement in type-ii parametric down-conversion. Phys Rev A 50(1):23
19. Babichev SA, Appel J, Lvovsky AI (2004) Homodyne tomography characterization and nonlocality of a dual-mode optical qubit. Phys Rev Lett 92(19):193601
20. Fuwa M, Takeda S, Zwierz M, Wiseman HM, Furusawa A (2015) Experimental proof of nonlocal wavefunction collapse for a single particle using homodyne measurements. Nat Commun 6
21. Hardy Lucien (1994) Nonlocality of a single photon revisited. Phys Rev Lett 73(17):2279
22. Hessmo Björn, Usachev Pavel, Heydari Hoshang, Björk Gunnar (2004) Experimental demonstration of single photon nonlocality. Phys Rev Lett 92(18):180401
23. Steve James Jones and Howard Mark Wiseman (2011) Nonlocality of a single photon: paths to an Einstein-Podolsky-Rosen-Steering experiment. Phys Rev A 84(1):012110
24. Morin Olivier, Bancal Jean-Daniel, Ho Melvyn, Sekatski Pavel, D'Auria Virginia, Gisin Nicolas, Laurat Julien, Sangouard Nicolas (2013) Witnessing trustworthy single-photon entanglement with local homodyne measurements. Phys Rev Lett 110(13):130401
25. Tan SM, Walls DF, Collett MJ (1991) Nonlocality of a single photon. Phys Rev Lett 66(3):252
26. Greenberger DM, Horne MA, Zeilinger A (1995) Nonlocality of a single photon? Phys Rev Lett 75(10):2064
27. Werner RF (1989) Quantum states with Einstein-Podolsky-Rosen correlations admitting a hidden-variable model. Phys Rev A 40(8):4277

28. Van Enk SJ (2005) Single-particle entanglement. Phys Rev A 72(6):064306
29. Drezet Aurélien (2006) Comment on "single-particle entanglement". Phys Rev A 74(2):026301
30. Pawłowski Marcin, Czachor Marek (2006) Degree of entanglement as a physically ill-posed problem: the case of entanglement with vacuum. Phys Rev A 73(4):042111
31. Van Enk SJ (2006) Reply to "comment on 'single-particle entanglement"'. Phys Rev A 74(2):026302
32. Beugnon J, Jones MP, Dingjan J, Darquié B, Messin G, Browaeys A, Grangier P (2006) Quantum interference between two single photons emitted by independently trapped atoms. Nature, 440(7085):779–782
33. Feng Xun-Li, Zhang Zhi-Ming, Li Xiang-Dong, Gong Shang-Qing, Zhi-Zhan Xu (2003) Entangling distant atoms by interference of polarized photons. Phys Rev Lett 90(21):217902
34. Wootters WK, Zurek WH (1982) A single quantum cannot be cloned. Nature, 299(5886):802–803
35. Boschi Danilo, Branca Salvatore, De Martini Francesco, Hardy Lucien, Popescu Sandu (1998) Experimental realization of teleporting an unknown pure quantum state via dual classical and Einstein-Podolsky-Rosen channels. Phys Rev Lett 80(6):1121
36. Bouwmeester Dik, Pan Jian-Wei, Mattle Klaus, Eibl Manfred, Weinfurter Harald, Zeilinger Anton (1997) Experimental quantum teleportation. Nature 390(6660):575–579
37. Vaidman Lev, Yoran Nadav (1999) Methods for reliable teleportation. Phys Rev A 59(1):116
38. DelRe E, Crosignani B, Di Porto P (2000) Scheme for total quantum teleportation. Phys Rev Lett 84(13):2989
39. Scully MO, Englert BG, Bednar CJ (1999) Two-photon scheme for detecting the bell basis using atomic coherence. Phys Rev Lett 83(21):4433
40. Garziano Luigi, Macrì Vincenzo, Stassi Roberto, Di Stefano Omar, Nori Franco, Savasta Salvatore (2016) One photon can simultaneously excite two or more atoms. Phys Rev Lett 117(4):043601
41. Asbóth JK, Calsamiglia J, Ritsch H (2005) Computable measure of nonclassicality for light. Phys Rev Lett 94(17):173602

Chapter 7
Super-Crystals in Composite Ferroelectrics

In this chapter, in Sect. 7.1, we report the discovery of spontaneous polarization "Super-crystals" in microstructured disordered ferroelectric samples, published in the article *'Super-crystals in composite ferroelectrics'*, Nature Communications **7**, 10674 (2016).

This phenomenon is observed during a paraelectric-ferroelectric phase transition of specifically grown potassium-lithium-tantalate-niobate (KLTN) crystals [1]. In contrast with a generic phase-transition that leads to a disordered mosaic of polar domains, we found a coherent three-dimensional domain structure of polarization that permeates the whole sample. Visible light propagation reveals a polarization "super-crystal" with a micrometric lattice constant, a phase that mimics standard solid-state structures but on scales that are thousands of times larger.

In Sect. 7.2 we investigate the evolution of the state of polarization of light propagating through this three-dimensional domain structure. In distinction to standard depoled ferroelectrics, where random birefringence causes depolarization and scattering, light is observed to suffer varying degrees of depolarization and remains fully polarized when linearly polarized along the crystal principal axes. The effect is found to be supported by the formation of polarized speckles organized into a spatial lattice and occurs as the ferroelectric settles into the spontaneous "super-crystal". The polarization lattices gradually disappear as the ferroelectric state reduces to a disordered distribution of polar-nano regions above the critical point. This second phenomenon has been published in *'Observation of polarization-maintaining light propagation in depoled compositionally disordered ferroelectrics'*, Optics Letters **42**, 3856 (2017).

7.1 Super-Crystals in Composite Ferroelectrics

As atoms and molecules condense to form solids, a crystalline state can emerge with its highly ordered geometry and subnanometric lattice constant. In some physical systems, such as ferroelectric perovskites, a perfect crystalline structure forms even when the condensing substances are non-stoichiometric. The resulting solids

G. Di Domenico, *Electro-optic Photonic Circuits*, Springer Theses,
https://doi.org/10.1007/978-3-030-23189-7_7

have compositional disorder and complex macroscopic properties, such as giant susceptibilities and non-ergodicity. Here we observe the spontaneous formation of a cubic structure in composite ferroelectric potassium-lithium-tantalate-niobate with micrometric lattice constant, 10^4 times larger than that of the underlying perovskite lattice. The 3D effect is observed in specifically designed samples in which the substitutional mixture varies periodically along one specific crystal axis. Laser propagation indicates a coherent polarization super-crystal that produces an optical x-ray diffractometry, an ordered mesoscopic state of matter with important implications for critical phenomena and applications in miniaturized three-dimensional optical technologies.

Textbook models of global symmetry breaking include a low-symmetry low-temperature state with a fixed infinitely extended coherence. In contrast, the spontaneous polarization observed as spatial inversion symmetry is broken during a paraelectric-ferroelectric phase-transition generally leads to a disordered mosaic of polar domains that permeate the finite samples [63]. Coherent and ordered ferroelectric states with remarkable properties of both fundamental and technological interest [15, 23, 29, 35] can emerge when ferroelectricity is influenced by external factors, such as system dimensionality [17], strain gradients [6, 10, 11], electrostatic coupling [8, 9] and magnetic interaction [5, 34]. We report the spontaneous formation of an extended coherent three-dimensional (3D) superlattice in the nominal ferroelectric phase of specifically grown potassium-lithium-tantalate-niobate (KLTN) crystals [19–21, 53]. Visible light propagation reveals a polarization super-crystal with a micrometric lattice constant, a counterintuitive mesoscopic phase that naturally mimics standard solid-state structures but on scales that are thousands of times larger. The phenomenon is achieved using compositionally disordered ferroelectrics [12, 24, 33, 44, 51, 52, 57, 69, 75, 76]. At one given temperature, these have the interesting property of a manifesting a single perovskite phase whose dielectric properties depend on the specific composition [37, 67, 81]. For example, a compositional gradient along the pull axis leads to a position-dependent Curie point $T_C(\mathbf{r})$, so that for a given value of crystal temperature T a phase separation occurs, with the regions where $T > T_C$ being paraelectric and those with $T < T_C$ developing a spontaneous polarization [74]. Specifically tailored growth schemes are even able to achieve an oscillating T_C along a given direction, say the x-axis [2, 18]. In these conditions, we can expect that at a given T in proximity of the average (macroscopic) T_C, the sample will be in a hybrid state with alternating regions with and without spontaneous polarization. Crossing the Curie point, in conditions in which perovskite polar-domains pervade the volume forming 90° configurations to minimize the free-energy associated to polarization charge [41], this oscillation can form a full three-dimensional periodic structure.

7.1.1 Experiments

Observation of a Compositionally-Induced Super-Crystal

To investigate the matter we make use of top-seeded ferroelectric crystals with an oscillating composition along the growth axis achieved using an off-center growth technique in the furnace [3, 18]. We obtain a zero-cut 2.4 mm by 2.0 mm by 1.7 mm, along x,y,z directions respectively, optical quality KLTN sample with a periodically oscillating niobium composition of period $\Lambda = 5.5\,\mu$m along the x-axis, with an average composition $K_{1-\alpha}Li_{\alpha}Ta_{1-\beta}Nb_{\beta}O_3$ with $\alpha = 0.04$ and $\beta = 0.38$ (see Methods). When the crystal is allowed to relax at $T = T_C - 2$ K, i.e., in proximity of the spatially averaged room-temperature Curie point $T_C = 294$ K, laser light propagating through the sample suffers relevant scattering with strongly anisotropic features (see Fig. 7.1a–d). Typical results are reported in Fig. 7.1b–d, and appear as an optical analogue of x-ray diffraction in low-temperature solids. This optical diffractometry provides basic evidence of a 3D superlattice at micrometric scales. Probing the principal crystal directions reveals several diffraction orders that map the entire reciprocal space. The large-scale super-crystal, that permeates the whole sample, overlaps—along the x-direction—with the built-in compositional oscillating seed (see Methods). The superlattice extends in full three-dimensions, with the same periodicity $\Lambda = 5.5\,\mu$m of the x-oriented compositional oscillation, also along the orthogonal y and z-directions. In particular, Fig. 7.1d indicates that in the plane perpendicular to the built-in dielectric microstructure Γ vector, i.e., where spatial symmetry should be

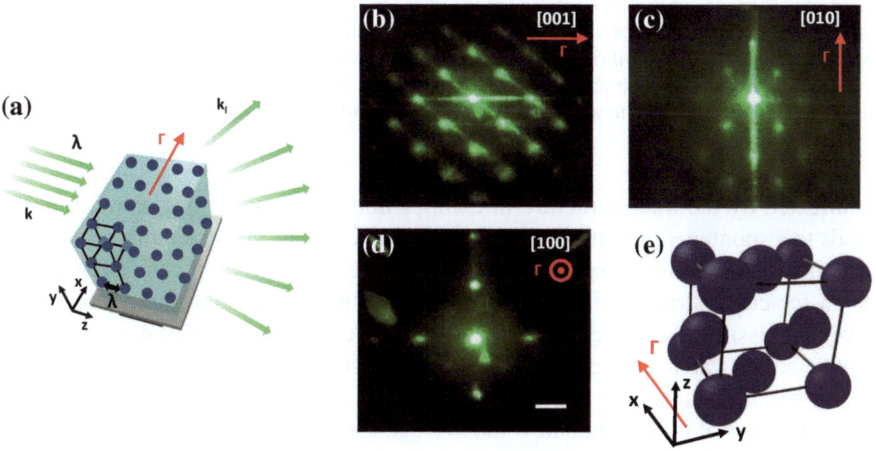

Fig. 7.1 Super-crystal in the ferroelectric phase. **a** Sketch of visible-light diffraction from micrometric structures through a transparent crystal and **b–d** 3D superlattice probed at $T = T_C - 2$ K along the principal symmetry direction of the crystal, respectively with the incident wavevector **k** parallel to **b** z-direction, **c** y-direction and **d** x-direction. Crystallographic analysis reveals the elementary cubic structure of lattice constant Λ shown in **e**. Scale bar corresponds to 1.2 cm

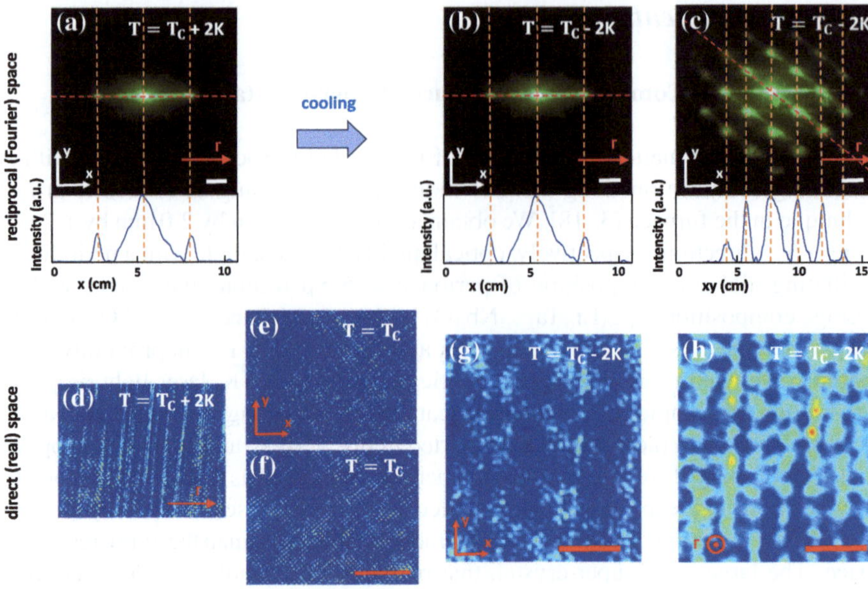

Fig. 7.2 Light diffraction in microstructured KLTN above and below the Curie point: observations in Fourier and real space. **a** Reciprocal space probed at $T = T_C + 2\,K$ (hot paraelectric phase), showing the first diffraction orders due to the one-dimensional sinusoidal compositional modulation. Cooling below the critical point results at $T = T_C - 2\,K$ (super-crystal ferroelectric phase) in **b** a supercooled (metastable) 1D superlattice with the same diffraction orders that relaxes at steady-state into **c** the cold (stable) super-crystals. Both in **b** and **c** the direction of incident light is othogonal to Γ, as in **a**. **d–h** Corresponding transmission microscopy images revealing **d** unscattered optical propagation, **e–f** scattering at the phase transition, **g** unscattered optical propagation in the metastable superlattice and **h** periodic intensity distribution underlining the 3D superlattice. Metastable and stable (equilibrium) phases are inspected respectively at times $t \approx 1$min and $t \approx 1$ h after the structural transition at $T = T_C$. Bottom profiles in **a–c** are extracted along the red dotted line. Scale bars correspond to **a–c** 1.2 cm, **d–f** 100 μm and **g–h** 10 μm

unaffected by the microstructure in composition, the ferroelectric phase-transition leads to a spontaneous pattern of transverse scale Λ. The corresponding elementary structure on micrometric spatial scales is reported in Fig. 7.1e; it can be represented as an fcc-cubic structure in which the occupation of one of the three faces ($z - y$ face) is missing [64]. The structure, which is, to our knowledge, not observed at atomic scales, can be reduced to an simple cubic structure with a three-fold basis and lattice parameter $a = \Lambda$.

As the crystal is brought below the average Curie point, it manifests a metastable (supercooled) and a stable (cold) phase, as analyzed in Fig. 7.2 both in the reciprocal (Fourier) and direct (real) space. In the nominal paraelectric phase, at $T = T_C + 2\,K$ (Fig. 7.2a), we observe the first Bragg diffraction orders (± 1) consistent with the presence of the seed microstructure, a one-dimensional transverse sinusoidal modulation acting as a diffraction grating; the distance from the central 0-order fulfills the Bragg condition,that is, scattered light forms an angle $\theta_B = \lambda/2n_0\Lambda \simeq 7°$ with the incident wavevector **k**. Crossing the ferroelectric phase-transition temperature T_C (see Methods) we detect a supercooled metastable state that has an apparently analogous diffraction effect (Fig. 7.2b) that is dynamically superseded by the stable and coherent cold superlattice phase (Fig. 7.2c), in which spatial correlations are extended to the whole crystal volume. In real space, transmission microscopy (see Methods) shows unscattered optical propagation through the paraelectric sample at $T = T_C + 2\,\text{K}$ (Fig. 7.2d), that turns into critical opalescence and scattering from obliques random domains at the structural phase transition (Fig. 7.2e–f), and in unscattered transmission in the metastable ferroelectric phase at $T = T_C - 2\,\text{K}$ (Fig. 7.2g). After dipolar relaxation has taken place, the cold super-crystal appears in this case as a periodic intensity distribution on micrometric scales, as shown in Fig. 7.2h.

Spontaneous Polarization Underlying the Ferroelectric Superlattice

To further analyze these supercooled and cold phases, we inspect the supercooled one-dimensional phase (Fig. 7.2b) that is accessible through linear (unbiased) and electro-optic (biased) polarization-resolved Bragg diffraction measurements. In particular, referring to the setup illustrated in Fig. 7.3a, we measure the diffraction efficiency $\eta = P_B/(P_B + P_0)$, where P_B and P_0 are respectively the diffracted and non-diffracted power, in the first Bragg resonance condition, i.e., with the incident wavevector **k** forming the angle θ_B respect to the z-axis. The diffraction efficiency η is reported in Fig. 7.3b for different input light polarization and temperature across the average Curie point. Diffraction strongly depends both on the nominal crystal phase and on the polarization of the incident wave: a large increase in η is found for light polarized in the x,z plane (H-polarized). For $T > T_C$ the dependence on light-polarization is consistent with what expected in standard periodically index-modulated media (wave-coupled theory), that is, a weak temperature dependence and a maximum η for light polarized normal to the grating vector (V-polarized). In this case, the difference in η_H ($\triangle$) and η_V ($\square$) can be related to the different Fresnel coefficient governing interlayer reflections and is congruently $\eta_V > \eta_H$ by an amount that decreases for larger θ_B [31, 80]. Consistently, the (H + V)-polarized curve ($\circ$), that is when the input linear polarization is at 45° with respect to the H and V polarizations, falls between these two curves. Standard behavior is violated for $T < T_C$, where a large enhancement in η_H rapidly leads to a regime with $\eta_V < \eta_H$.

The physical underpinnings of the super-crystal can be grasped considering the simple model illustrated in Fig. 7.3c. Here we consider the metastable 1D superlattice (Fig. 7.2b) before tensorial effects cause the full 3D superlattice relaxation (Fig. 7.2c). Specifically, for a given T, regions with a local value of T_C such that $T < T_C$ (dark shading) will manifest a finite spontaneous polarization $P_S \neq 0$, whereas region with $T > T_C$ (light shading) will have a $P_S \simeq 0$. Optical measurements are sensitive to the square of the crystal polarization $\langle \mathbf{P} \cdot \mathbf{P} \rangle \simeq P_S^2$ through the resulting index pattern modulated via the quadratic elecro-optic response $\delta n(P) = -(1/2)n^3 g P^2$, where n is the unperturbed refraction index and g is the proper perovskite elecro-

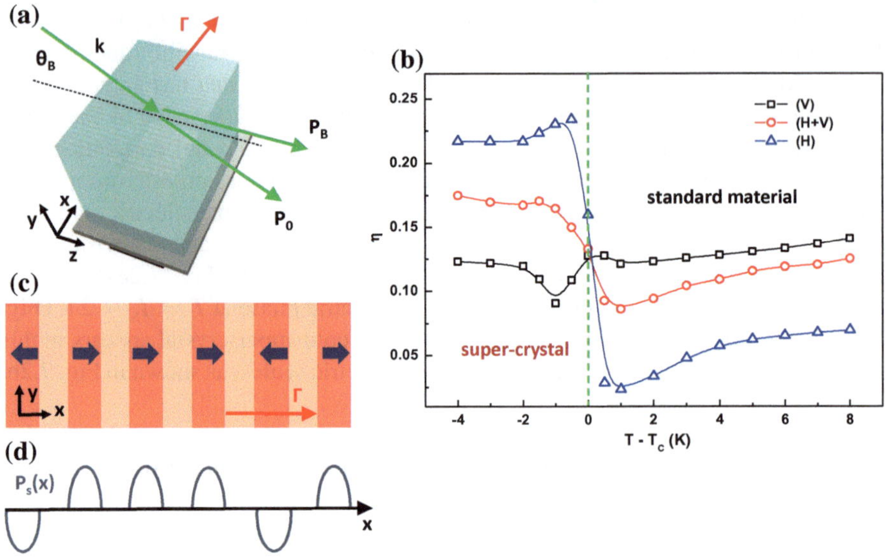

Fig. 7.3 Diffractive behaviour of the 1D supercooled superlattice. **a** Sketch of the experimental geometry and **b** detected diffraction efficiency (dots) as a function of temperature in proximity of ferroelectric transition for different wave polarizations. An anomaly appears crossing T_C for H-polarized light signaling the emergence of the super-crystal. Lines are interpolations serving as guidelines. **c** Scheme of the periodically-ordered ferroelectric state along the x-direction undelying the super-crystal phase for $T < T_C$ and giving the spontaneous polarization $P_S(x)$ sketched in panel **d**

optic coefficient [7, 52]. Enhanced Bragg-scattering of light polarized parallel to the seed direction Γ (H in Fig. 7.3b—super-crystal) indicates that $P_S(x)$ is parallel to the seed direction (x-axis), where the elecro-optic coefficient g have its maximum value $g = 0.16\text{m}^4/\text{C}^2$. The resonant response at θ_B and the absence of higher harmonics (Fig. 7.2b) indicates that this $P_S(x)^2$ distribution is sinusoidal with wavevector Γ. Hence, although in general it may be that macroscopically $\langle \mathbf{P} \rangle \simeq 0$, it turns out that $\langle \mathbf{P}^2 \rangle \simeq P_S^2 \neq \langle \mathbf{P} \rangle^2 \neq 0$ on the micrometric scales, in analogy with optical response in crystals affected by polar-nanoregions [25, 52, 76]. Optical diffraction efficiency reported in Fig. 7.3b then occurs considering $\eta = \sin^2\left(\frac{\pi d(\delta n)}{\lambda \cos\theta_B}\right)$, with resonant enhanced diffraction for $T < T_C$ caused by $\delta n = \delta n_0 + \delta n(P)$, where $\delta n_0 \sim 10^{-4}$ is the polarization independent index change due to the periodic composition variation (Sellmeier's index change).

Electro-Optical Diffraction Analysis

To validate this picture we perform electro-optic diffractometry experiments, in which a macroscopic polarization activating the nonlinear periodic response is

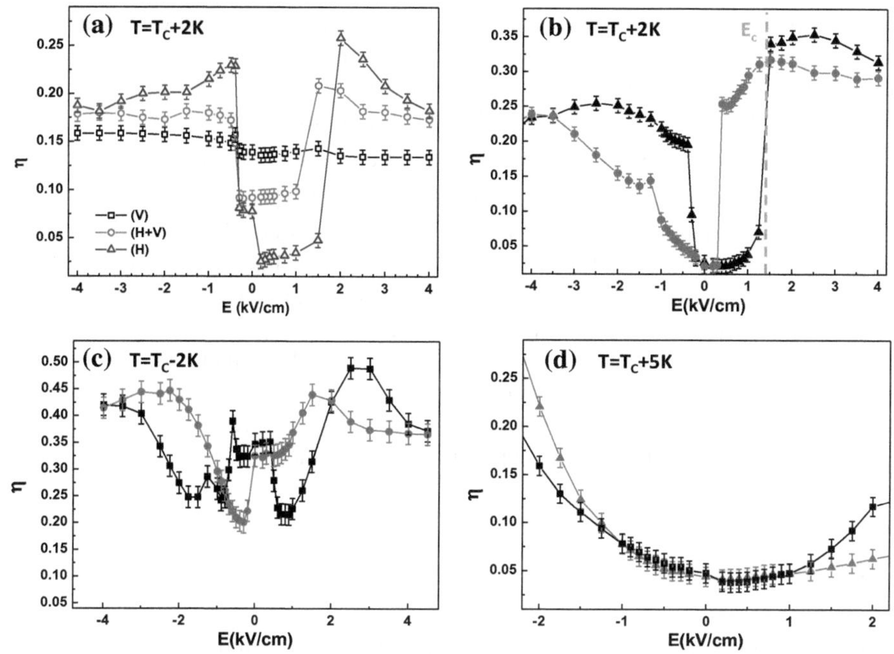

Fig. 7.4 Electro-optic Bragg diffraction in the critical region. **a** Diffraction efficiency as a function of the external applied field for different light polarization at $T = T_C + 2\,\text{K}$; **b** hysteresis loop at the same temperature and **c** at $T = T_C - 2\,\text{K}$ for H-polarization. **d** Expected [2] weak-histeretic paraelectric (parabolic) behaviour at $T = T_C + 5\,\text{K}$. In **b**–**d** black dots and red dots indicates data obtained respectively increasing and decreasing the bias field. Lines are interpolations serving as guidelines. Error bars are given by the statistics on five experimental realizations

induced via an external static field E applied along x. Results are reported in Fig. 7.4; in particular, in Fig. 7.4a the polarization and field dependence of η is shown at $T = T_C + 2\,\text{K}$. We observe a nearly field-independent behavior for V-polarized light, that arises from its low electro-optic coupling (bias field and light polarization are orthogonal, $g = -0.02\text{m}^4/\text{C}^2$); differently, η_H increases with the field showing a "discontinuity" at the critical field $E_C =(1.4 \pm 0.1)$ kV/cm. The strong similarity between this enhancement and those observed in unbiased conditions at T_C (Fig. 7.3b) indicates that E_C coincides with the coercive field and the discontinuity corresponds to the field-induced phase-transition [3, 53, 75]. In fact, in Fig. 7.4b we repeat this experiment enhancing the experimental field-sensitivity and acquiring data also for decreasing field amplitudes. The result is a partial-hysteretic loop for the diffraction efficiency that demonstrates the field-induced transition and underlines that, both in the linear and nonlinear (electro-optic) case, the effect of the seeded ferroelectric ordering is to provide a periodic spontaneous polarization along x. We also note a slight asymmetry with respect to positive/negative fields; this is associated to a residual fixed space-charge field that may play an important role in the spontaneous polarization alignment process and hence in leading to a residual $\langle \mathbf{P} \rangle \neq 0$. The existence of a periodic spontaneous polarization distribution in the superlattice (Fig. 7.3c) is confirmed in Fig. 7.4c, where electro-optic Bragg diffraction below T_C is reported. An oscillating full-hysteretic behavior is observed as a function of the external field, consistently with the prediction $\eta(E) = \sin^2\left(\frac{\pi d(\delta n(E))}{\lambda \cos\theta_B}\right)$ with $\delta n(E) = \delta n_0 + (1/2)n^3 g(P_S^2 + 2\varepsilon_0\chi\langle P_S\rangle E + \varepsilon_0^2\chi^2 E^2)$. The increase in η due to the superlattice polarization allows us to explore its full sinusoidal behavior, that usually requires extremely large fields in the paraelectric phase and reduces to a parabolic behaviour (Fig. 7.4d) [79]. From this parabolic behaviour detected at $T = T_C + 5\,\mathrm{K}$ we estimate that the resulting amplitude in the point-dependent Curie temperature due to the compositional modulation is $\Delta T_C \simeq 2\,\mathrm{K}$ [2]. Agreement with the periodic polarization model is further stressed by deviations emerging in $\eta(E)$ especially for low and negative increasing fields, where the dependence on $\langle P_S\rangle$ make observations weakly dependent on the specific experimental realization.

7.1.2 Discussion

An interesting point arising from the experimental results and analysis is how the periodically-ordered polarization state along the x-direction leads to the super-crystal. Since we pass spontaneously from a metastable to a stable mesoscopic phase, polar-domain dynamics in presence of the fixed spatial scale Λ play a key role. In fact, we note that the 1D superlattice sketched in Fig. 7.3c involves the appearance of charge-density and associated strains between polar planes, so that the ferroelectric crystal naturally tends to relax into a more stable configuration. In standard perovskites, equilibrium configurations are mainly those involving a 180° and 90° orientation between adjacent polar domains, as schematically shown in Fig. 7.5a. To explain the 3D polar-state and its periodical features underlying the super-crystal, we consider the 90° configuration, which is characterized by 45° domain walls that we observe in a disordered configuration during the ferroelectric phase transition at T_C (Fig. 7.2f). Due to the periodic constraint along the x-axis, this arrangement has the unique property of reproducing our observations, minimizing energy associated to internal charge-density, and transferring the built-in 1D order to the whole volume with the same spatial scale Λ. We illustrate the domain pattern in Fig. 7.5b for the $x-y$ plane, whereas in Fig. 7.5c the elementary cell is shown in the three-dimensional case, where it maintains its stability features in terms of charge-density energy. In particular, in Fig. 7.5b, domain walls resulting in the diffraction orders of Fig. 7.1b are marked, as well as the 45° correlation period, that agree with optical observations of the reciprocal space. We further stress that vertical domains (light blue in Fig. 7.5b) are optically analoguous to paraelectric regions; moreover, 180° rotations in the polarization direction in each polar region has no effect on the optical response. In view of the symmetry of this arrangement, the observed diffrac-

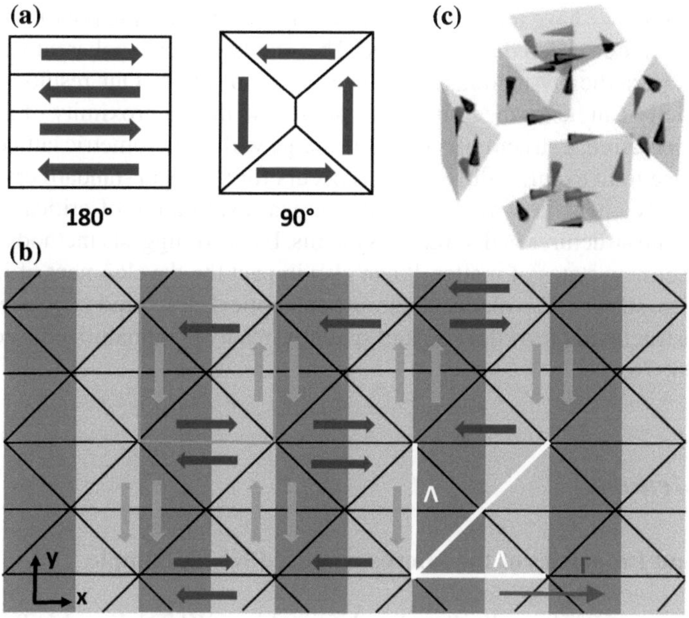

Fig. 7.5 Polar-domain configuration underlying the 3D superlattice. **a** Typical 180° and 90° domain configurations in perovskites ferroelectrics. **b** Planar domain arrangement scheme in the stable super-crystal phase obtained with elementary blocks of 90° configurations (green cell). In this periodically-ordered ferroelectric state the compositional modulation (as for Fig. 7.3c), other domain walls ruling optical diffractometry (black lines), and periods along x,y and xy-axis (white bars) are highlighted. Vertical polarizations have a lighter color to stress their weak optical response in our KLTN sample. **c** Extension of the single unit-cell (green cell in **b**) in three dimensions

tion anisotropy (Fig. 7.1d) is then associated to the absence of grating-planes in the $y - z$ face.

Further insight on the 3D domain structure requires numerical simulations based on Monte Carlo methods [36, 60] and phase-field models [14, 16, 39, 40]; they may confirm our picture and reveal new aspects for ferroelectricity, such as polar dynamics, spontaneous long-range ordering and the role of polar strains in composite ferroelectrics with built-in compositional microstructures. In fact, the effect of the composition profile is here crucial in triggering the spontaneous formation of the macroscopic coherent structure, as it sets the typical domain size along the x-direction and so rules the whole dynamic towards the equilibrium state. We expect that a different amplitude and period of the modulation may affect the formation, stability, time- and temperature-dynamics of the super-crystal; indeed, these parameters of the compositional gradient may be important in determining the interaction between polar-regions. Advanced growth techniques [2] can open future perspectives in this direction, as well as towards composite ferroelectrics with different compositional shapes of fundamental and applicative interest.

To conclude, we have reported the formation of a mesoscopic polarization supercrystal in a nanodisordered sample of KLTN. The large-scale coherent state is triggered by a periodically-modulated change in composition. Our results show how ferroelectricity can be arranged into new phases, so that in proximity of an average critical temperature, a structural order can emerge with a micrometric lattice constant so as to cause light to suffer diffraction as occurs for x-rays in standard crystals. The effect not only opens new avenues in the optical exploration of critical properties and large scale structures in disordered systems, but also suggests methods to predict and engineer new states of matter. It can also impact the development of innovative technologies, such as nonvolatile electronic and optical structured memories [15, 23, 35], microstructured piezo-devices and spatially-resolved miniaturized electro-optic devices [48, 76, 79].

7.1.3 Methods

Growth and Properties of the Microstructured KLTN Sample

We consider a compositionally disordered perovskite of KLTN, $K_{1-\alpha}Li_{\alpha}Ta_{1-\beta}Nb_{\beta}O_3$ with $\alpha = 0.04$ and $\beta = 0.38$, grown through the top-seeded solution method by extracting a zero-cut 2.4 mm by 2.0 mm by 1.7 mm, along x, y, z directions respectively, optical quality specimen. It shows, through low-frequency dielectric spectroscopy measurements, the spatial-averaged Curie point, which signals the transition from the high-temperature symmetric paraelectric phase to the low-temperature ferroelectric phase, at the room-temperature $T_C = 294$ K. A one-dimensional seed microstructure is embedded into the sample as it is grown through the off-center growth technique so as manifest a sinusoidal variation in the low-frequency dielectric constant, and thus in the critical temperature T_C, along the growth-axis (x-direction) [3, 18]. This dielectric volume microstructure causes an index of refraction oscillation of period $\Lambda = 5.5\,\mu$m, that is able to diffract light linearly and electro-optically [54]. Details on the technique employed in the sample growth can be found in Ref. [18]. We note here that the composition amplitude of the periodic microstructure can be estimated from $\Delta\beta/\Delta T$, where $\Delta\beta$ is the amplitude variation in niobium composition and ΔT is the change in the growth temperature incurred by the off-center rotation. At the growth temperature of approximatively 1470 K, the ratio $\Delta\beta/\Delta T \approx 0.35$ ‰·mol/K has been extracted from the phase diagram of KTN. The temperature variation incurred by the off-center rotation was measured to be 3 K, from which we obtain $\Delta\beta \approx 1.05$ ‰·mol.

Optical-Diffraction Experiments

The macroscopic linear and electro-optic diffractive properties of the crystal have been investigated launching low-power (mW) plane waves at $\lambda = 532$ nm that propagate normal and parallel to the grating vector Γ ($\Gamma = 2\pi/\Lambda$), that is along the x-direction (Fig. 7.3a). Light diffracted by the medium is detected using a broad-area charge-coupled-device (CCD) camera placed at $d = 0.2$ m from the crystal output facet or collected into Si power-meters. In real-space measurements (Fig. 7.2d–h) the output crystal facet is imaged on the CCD camera and a cross-polarizers setup [52, 76] has been used to highlight contrast due to polarization inhomogeneities. The time needed to obtain a fully-correlated state corresponding to the 3D super-crystal depends on the cooling rate τ and on the details of the thermal environment. Considering, for instance, as a thermal protocol a cooling rate $\tau = 0.05$ K/s and an environment at $T = T_C + 1$ K (weak thermal gradients), we have found that the metastable 1D lattice state at $T = T_C - 2$ K (Fig. 7.2b), in which correlations involve mainly the direction including the Γ vector, lasts approximatively 1 h. In this stage, although no macroscopic order occurs in the other directions [32], we observe optimal optical transmission of the sample (Fig. 7.2g); output light is not affected by scattering related to the existence of random domains and this underlines the presence of a mesoscopic ordering process in which the typical domain size is set. As regards the inspected temperature range, we have found that the super-crystal forms for temperatures till to $T = 288$ K, although correlations are weaker at the lower temperatures. This is consistent with the fact that at these temperatures also the regions with a lower local T_C are well below the transition point.

7.2 Observation of Polarization-Maintaining Light Propagation in Depoled Compositionally Disordered Ferroelectrics

The polarization of light is strongly affected by anisotropy, the paradigm being the birefringence observed in non-centrosymmetric crystals, such as ferroelectrics [63], where the index of refraction depends on the polarization and propagation direction. In a full three-dimensional scenario, birefringence can also affect wave propagation, not only introducing coherent scattering [7], but even engendering localized guided modes. More exotic polarization phenomena can be observed in systems where the anisotropy emerges on periodic structures, as occurs in metamaterials [46, 47, 59, 65], in periodically-poled multiferroic media [28] and in organic ferroelectrics [38]. Elaborate anisotropic states are also found in nanodisordered ferroelectrics in proximity of the Curie point signaling the paraelectric-ferroelectric phase transition, where so-called polar-nanoregions (PNRs) emerge [27, 43, 58, 61, 77, 83]. These are known to form the basis for remarkable optical responses of both fundamental and technological interest, such as randomly-matched second harmonic generation [4, 66, 78], the giant photorefractive nonlinearities [19, 20, 22, 42, 53, 55], giant quadratic electro-optic coefficients [12, 25, 52, 71, 73], strong electromechanical coupling [45] and the anomalous electro-optic effect [52, 72, 76]. Polar-nanoregions generally form a disordered three-dimensional mosaic for which optical birefringence experiments indicate average local symmetry breaking, providing a tool for the study of possible non-ergodic behavior and dipolar-glass dielectric relaxation [30, 62, 70, 82]. In turn, the presence of large polar domains below the Curie point or in the paraelectric phase under high electric field [25], causes complete depolarization of propagating optical fields, a result of multiple interference of random scattered waves. This complicates the use of giant ferroelectric response in photonic applications, whether these be based on quasi-phase-matching and nonlinear electronic susceptibility or simply the electro-optic response [13, 26, 48, 74, 85]. In recent studies, conditions have been found in which the polar-nanoregion mosaic spontaneously settles into a three-dimensional coherent and periodic structure, a ferroelectric super-crystal with intriguing optical diffraction properties [56]. In this case, interesting effects are expected to emerge in light-polarization dynamics from the interplay of mesoscopic domain ordering and anisotropy, all this in a volume scenario.

We investigate the evolution of the optical polarization state in a super-crystal supported by nanodisordered KNTN and KLTN ferroelectrics. Here, spontaneous polarization leads to a periodic three-dimensional polar lattice with strong inhomogeneity and anisotropy at the micrometer scale. Polarimetric experiments indicate that, in distinction to pure depoled ferroelectrics, light propagates fully polarized for a linear polarization along the super-crystal principal axes. Analyzing the wave spatial distribution, we found the effect to be associated with the formation of speckle distributions of alternating orthogonally polarized states that spatially separate the input polarization into its linear polarization components parallel to the principal super-crystal axes. Precursors of the phenomenon are observed also above the Curie point, where the ferroelectric super-crystal vanishes, thus indicating preferred orientations of polar-nanoregions.

7.2.1 Experiments

To grasp the physics underlying the polarization dynamics reported, we note that a three-dimensional disordered mosaic of anisotropic domains, as schematically shown in Fig. 7.6a, dephases the components of an incident optical field according to the local random optical axis, so as to scatter and depolarize the transmitted light irrespective of its input polarization state. In turn, an ordered volume pattern of polar domains, such as the one encountered in a ferroelectric super-crystal and illustrated in Fig. 7.6b, where each lattice cell may host a so-called Kittel-like vortex [56], leads to a qualitatively different polarization evolution. In fact, in this case ferroelectric anisotropy is inhomogeneous but the local optical axis is on average aligned along the super-crystal structure.

In our experiments we used super-crystals formed in two different composite ferroelectrics to demonstrate the generality of the polarization-maintaining scenario. The KNTN super-crystal was obtained allowing a $K_\alpha Na_{1-\alpha}Ta_\beta Nb_{1-\beta}O_3$ ($\alpha = 0.85$, $\beta = 0.63$) solid-solution crystal to equilibrate for approximately 30 minutes below its

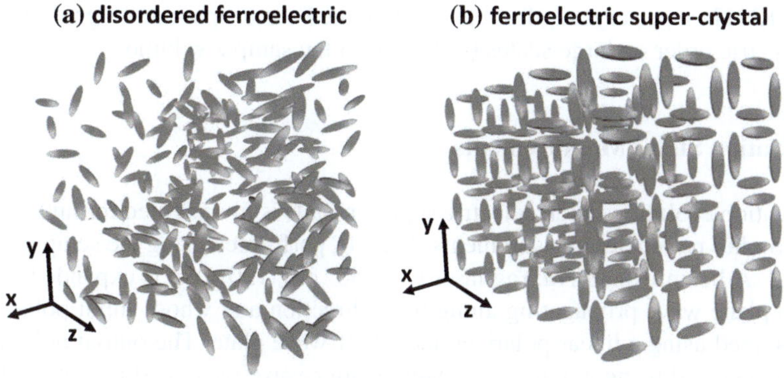

Fig. 7.6 Schematic of different spontaneous ferroelectric states. **a** Three-dimensional disordered distribution of polar domains. **b** Example of a volume polar structure underlying a ferroelectric super-crystal

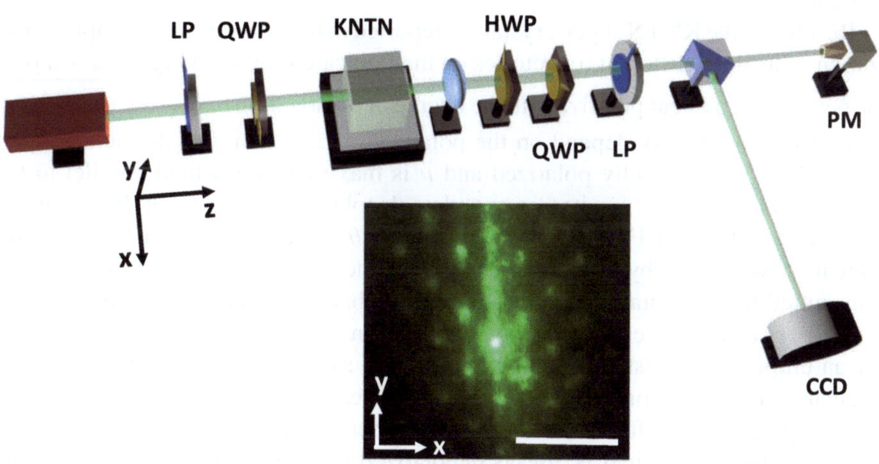

Fig. 7.7 Sketch of the experimental setup: linear polarizer (LP), quarter-wave plate (QWP), half-wave plate (HWP), power meter (PM) and imaging camera (CCD). The inset shows the optical diffraction pattern of the ferroelectric super-crystal embedded in the KNTN sample. Scale bar is 5 mm

ferroelectric-paraelectric Curie point $T = T_c - 3\,\text{K}$, with $T_c = 293.5\,\text{K}$. The crystal is grown through the top-seeded solution method by extracting a zero-cut optical-quality 2.1 mm by 2.5 mm by 2.6 mm specimen (along the x,y,z axes). The KLTN super-crystal emerged from the equilibration of a 2.4 mm by 2.0 mm by 1.58 mm $K_{\alpha}Li_{1-\alpha}Ta_{\beta}Nb_{1-\beta}O_3$ ($\alpha = 0.96$, $\beta = 0.62$) with $T_c = 294\,\text{K}$. Dielectric response for both samples is detailed in Refs. [49, 50] and the super-crystal formation process is reported in Ref. [56]. In Fig. 7.7 we show the optical diffraction pattern observed for the KNTN super-crystal: discrete spots fill the Fourier space and signal a periodic micrometric order on large scales ($\approx$25 μm) in the sample volume.

Polarization States Measurements

Polarization evolution in ferroelectric super-crystals is investigated through conventional Stokes parameter measurements [68, 84], performed using the setup shown in Fig. 7.7. A beam from a Nd:Yag laser ($\lambda = 532\,\text{nm}$, $150\,\text{mW}$) is expanded so as to form a plane wave propagating along the z direction and whose input polarization state is fixed using a linear polarizer and a half-wave plate. The output polarization state is analyzed using a half-wave plate, quarter-wave plate, and a linear polarizer placed after the sample. This allows the decomposition of the field into its Stokes components, i.e., horizontal (parallel to the x axis) S_H, 45 degrees S_{45} and right-circular S_R from the optical intensity detected through a power meter and a CCD camera.

Results for the KNTN super-crystal are reported in Fig. 7.8 varying the input polarization state along different trajectories on the Poincaré sphere. In Fig. 7.8a is shown the behavior of a linear polarization; the degree of polarization $\nu = \sqrt{S_H^2 + S_{45}^2 + S_R^2}$ is observed to strongly depend on the polarization direction θ of the field. In particular, light remains fully polarized and ν is maximum for a field parallel to the x or y axis ($\theta \simeq 0, 180°$), whereas complete depolarization occurs in the conjugate points ($\theta \simeq 90, 270°$). For intermediate values of θ, evolution of the horizontal component is well fitted by $S_H(\theta) = \cos\theta$ (blue line in Fig. 7.8a, b). Moreover, the transmitted light maintains a polarized fraction that almost coincides with S_H, that is, $\nu(\theta) = |\cos\theta|$ (green line in Fig. 7.8a, b). An analogous evolution is observed for an elliptical input state oriented along the x axis (Fig. 7.8b). In this case, the circular components completely depolarize whereas the linear horizontal and vertical ones propagate unaffected in the spatially inhomogeneous ferroelectric structure. Moreover, the output field is always depolarized ($\nu \simeq 0$) along a trajectory on the Poincaré sphere orthogonal to the H axis (Fig. 7.8c). The whole picture is observed in both super-crystals, is found to be independent of the laser wavelength (532–633 nm) and crystal orientation, and occurs equally for light propagating along the x and y directions of the sample. This suggest a strong connection between the observed depolarization and the one reported in electro-optic experiments in similar crystals in proximity of T_c under large electric field [12, 25, 42, 52], an effect that has been only partially understood.

Analysis in Near-Field

To pinpoint the underlying mechanism we perform spatially-resolved Stokes parameter maps of the transmitted light. In Fig. 7.9a we report the detected S_H for the significant case of a linearly polarized ($\theta = 90°$ in Fig. 7.8a) input wave from a

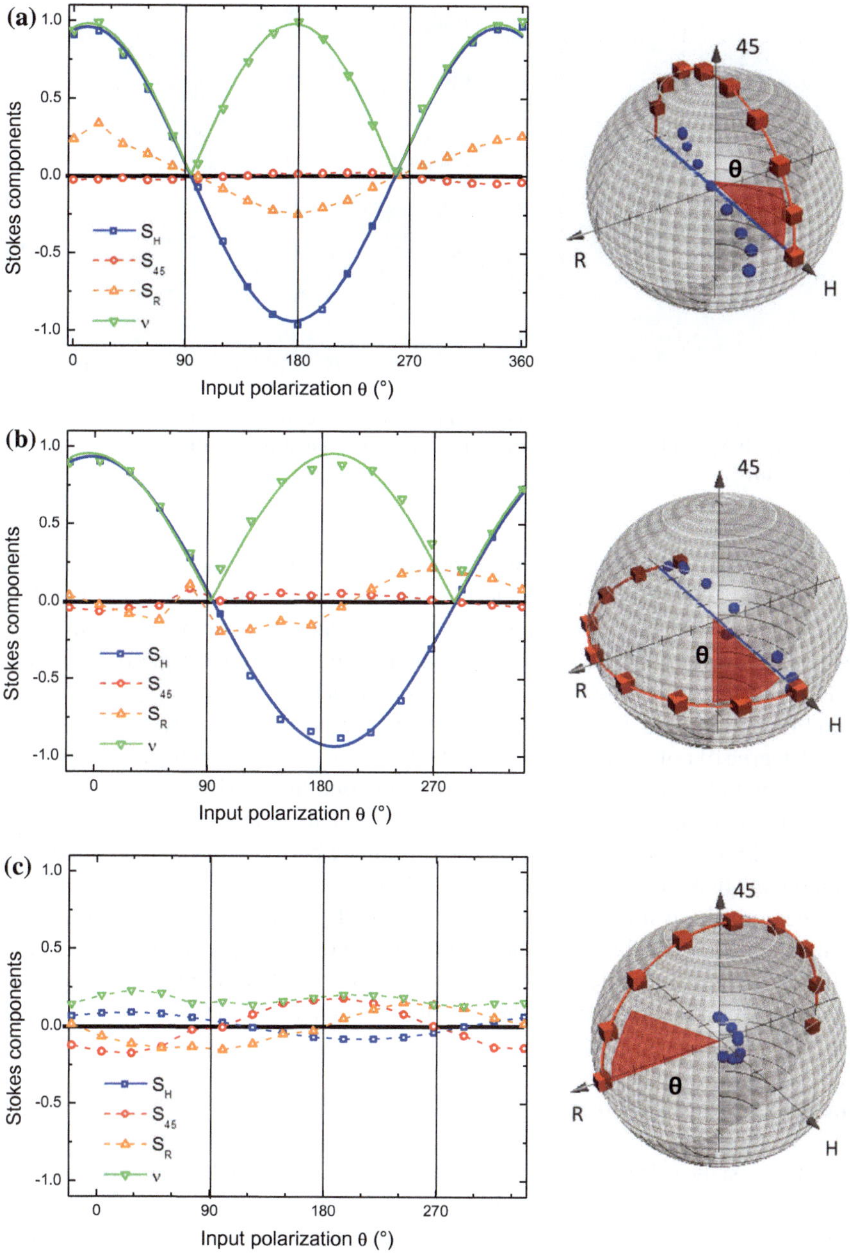

Fig. 7.8 Light-polarization dynamics in the KNTN super-crystal. **a–c** Stokes parameters measured varying the input polarization state along equators of the Poincaré sphere through the angular coordinate θ. Blue squares indicate the horizontal component of the polarization, red circles the 45° one, orange and green triangles are, respectively, the right-handed part and the degree of polarization ν. Solid lines are fitting functions (see main text) and dashed lines serves as guides. Insets show the corresponding input (red cubes) and output (blue spheres) states in the Poincaré space

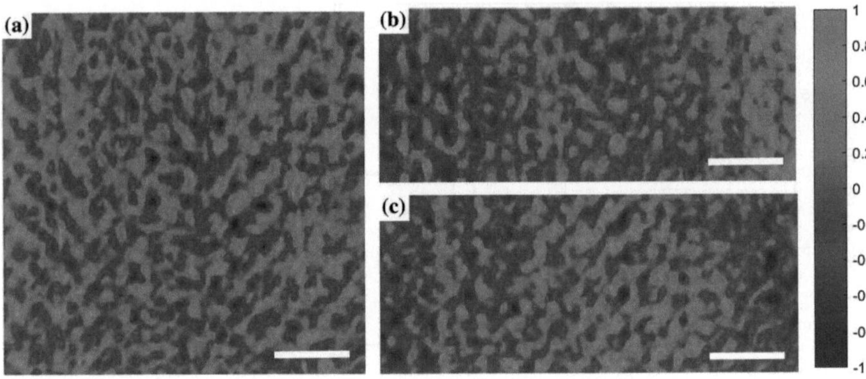

Fig. 7.9 Evidence of a locally-polarized speckle lattice. Stokes parameter maps showing the transmitted **a** S_H **b** S_{45} and **c** S_R local components for a linear input polarization with $\theta = 90°$. Scale bar is 20 μm

He-Ne laser ($\lambda = 633$ nm, 15 mW) propagating along the z direction of the KLTN super-crystal. We observe a speckle-like distribution arising from scattering during propagation. However, in contrast to what is expected for depolarized light from scattering, speckles distribute on a periodic lattice with approximately 6 μm lattice constant that coincides with the super-crystal structure. A similar speckle lattice is found for the S_{45} and S_R map, Fig. 7.9b and c, respectively. Interestingly, while the global degree of polarization is $\nu \simeq 0$, the degree of polarization measured averaging on each spatial point is $\nu \simeq 0.7$. Therefore, the output polarization state consists of a mixture of spatially-separated polarized states. This indicates how inhomogeneity of the medium introduces a local phase difference between orthogonal polarization components that strongly varies in space. A macroscopic Stokes measure (Fig. 7.7) averages out these local phases so that the field appears as depolarized although horizontal and vertical components are maintained during propagation. The optical polarization lattice closely follows the super-crystal, and this demonstrate a principal role played by the underlying ferroelectric state (see Fig. 7.6b).

The Role of Phase Transition

To further test the role of ferroelectric domains, we perform polarimetric transmission experiments varying the crystal temperature, so as to introduce strong fluctuations in super-crystal order, ultimately crossing the Curie temperature to restore global inversion symmetry in the paraelectric phase. As reported in Fig. 7.10a for the dynamics of a linearly polarized input state in KNTN (see Fig. 7.8a for a comparison), the average depolarization of the input wave appears less pronounced at $T = T_c + 6$ K. Although ν still depends on θ, its minimum at $\theta \approx 90°$ no longer vanishes. Although the main Stokes component remains S_H, other components become significant. Polarization evolution now is found to depend on the length of the sample along the propagation

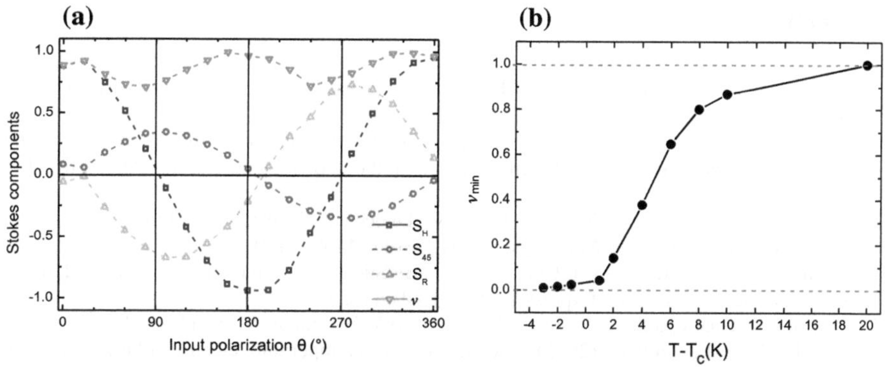

Fig. 7.10 Partial depolarization from polar-nanoregions. **a** Stokes parameters measured in the nominal paraelectric phase ($T = T_c + 6K$) for propagation along the x direction of the KNTN crystal. Inputs are linearly polarized, that is, θ varies along the equator of the sphere in Fig. 7.8a. **b** Minimum degree of polarization ν_{min} for measurements as in **a** versus $T - T_c$

axis and on λ, suggesting a macroscopic birefringence of the hosting paraelectric medium. This is consistent with the fact that above the Curie point the super-crystal is superseded by a disordered distribution of PNRs [45, 52, 71, 72, 76] that acts as a precursor of the macroscopic phase transition, so that no macroscopic index of refraction periodic lattice emerges. As reported in Fig. 7.10b, the corresponding minimum degree of polarization (ν_{min}) increases as temperature moves away from the critical point, an order parameter further underlining the role of the ferroelectric inhomogeneous structure. At $T = T_c + 20\,\text{K}$, where the crystal appears no longer affected by polar-nanoregions (proper paraelectric phase), $\nu = 1$ for all input polarizations.

7.2.2 Conclusion

To conclude, we have experimentally investigated optical polarization dynamics in the ferroelectric phase of compositionally-substituted perovskite ferroelectrics, in conditions for which an ordered three-dimensional polar-lattice is embedded in the material. The interplay between ferroelectric anisotropy and inhomogeneity leads to a new scenario in which the propagating wave can maintain its degree of polarization. Specifically, the field is separated into its linearly polarized components parallel to the super-crystal principal axes in the form of two spatially distinct periodic speckle patterns. Experiments above the Curie temperature suggest that polar nanoregions have preferred orientations along the crystal axes, a fact that may play a crucial role in phenomena involving the giant electro-optic and giant piezoelectric effect. Our results demonstrate how ordered polar domains can coherently modify the polarization of light, possibly enabling the use of the unconventional ferroelectric properties in photonic applications based on polarization control.

References

1. Agranat A, Hofmeister R, Yariv A (1992) Characterization of a new photorefractive material: K 1- ylyt 1- xnx. Opt Lett 17(10):713–715
2. Agranat AJ, Kaner R, Perpelitsa G, Garcia Y (2007) Stable electro-optic striation grating produced by programed periodic modulation of the growth temperature. Appl Phys Lett 90(19):192902
3. Agranat AJ, de Oliveira CEM, Orr G (2007) Dielectric electrooptic gratings in potassium lithium tantalate niobate. J Non-Cryst Solids 353(47):4405–4410
4. Ayoub M, Imbrock J, Denz C (2011) Second harmonic generation in multi-domain χ 2 media: from disorder to order. Opt Express 19(12):11340–11354
5. Baledent V, Chattopadhyay S, Fertey P, Lepetit MB, Greenblatt M, Wanklyn B, Saouma FO, Jang JI, Foury-Leylekian P (2015) Evidence for room temperature electric polarization in r m n 2 o 5 multiferroics. Phys Rev Lett 114(11):117601
6. Biancoli A, Fancher CM, Jones JL, Damjanovic D (2015) Breaking of macroscopic centric symmetry in paraelectric phases of ferroelectric materials and implications for flexoelectricity. Nat Mater 14(2):224–229
7. Bitman A, Sapiens N, Secundo L, Agranat AJ, Bartal G, Segev M (2006) Electroholographic tunable volume grating in the g 44 configuration. Opt Lett 31(19):2849–2851
8. Bousquet E, Dawber M, Stucki N, Lichtensteiger C, Hermet P, Gariglio S, Triscone J-M, Ghosez P (2008) Improper ferroelectricity in perovskite oxide artificial superlattices. Nature 452(7188):732–736
9. Callori SJ, Gabel J, Dong S, Sinsheimer J, Fernandez-Serra MV, Dawber M (2012) Ferroelectric pbtio 3/srruo 3 superlattices with broken inversion symmetry. Phys Rev Lett 109(6):067601
10. Catalan G, Janssens A, Gijsbert Rispens S, Csiszar OS, Rijnders G, Blank DHA, Noheda B (2006) Polar domains in lead titanate films under tensile strain. Phys Rev Lett 96(12):127602
11. Gustau Catalan A, Lubk AHGV, Snoeck E, Magen C, Janssens A, Rispens G, Rijnders G, Blank DHA, Noheda B (2011) Flexoelectric rotation of polarization in ferroelectric thin films. Nat Mater 10(12):963–967
12. Chang Y-C, Wang C, Yin S, Hoffman RC, Mott AG (2013) Giant electro-optic effect in nanodisordered ktn crystals. Opt Lett 38(22):4574–4577
13. Chao J-H, Zhu W, Chen C-J, Campbell AL, Henry MG, Yin S, Hoffman RC (2017) High speed non-mechanical two-dimensional ktn beam deflector enabled by space charge and temperature gradient deflection. Opt Express 25(13):15481–15492
14. Chen L-Q (2008) Phase-field method of phase transitions/domain structures in ferroelectric thin films: a review. J Am Ceram Soc 91(6):1835–1844
15. Choi KJ, Biegalski M, Li YL, Sharan A, Schubert J, Uecker R, Reiche P, Chen YB, Pan XQ, Gopalan V et al (2004) Enhancement of ferroelectricity in strained batio3 thin films. Science 306(5698):1005–1009
16. Chu P, Chen DP, Wang YL, Xie YL, Yan ZB, Wan JG, Liu J-M, Li JY (2014) Kinetics of 90 domain wall motions and high frequency mesoscopic dielectric response in strained ferroelectrics: a phase-field simulation. Sci Rep 4:5007
17. Dawber M, Rabe KM, Scott JF (2005) Physics of thin-film ferroelectric oxides. Rev Mod Phys 77(4):1083
18. De Oliveira CEM, Orr G, Axelrold N, Agranat AJ (2004) Controlled composition modulation in potassium lithium tantalate niobate crystals grown by off-centered tssg method. J Cryst Growth 273(1):203–206
19. DelRe E, Spinozzi E, Agranat AJ, Conti C (2011) Scale-free optics and diffractionless waves in nanodisordered ferroelectrics. Nat Photonics 5(1):39–42
20. DelRe E, Di Mei F, Parravicini J, Parravicini G, Agranat AJ, Conti C (2015) Subwavelength anti-diffracting beams propagating over more than 1,000 rayleigh lengths. Nat Photonics
21. Di Mei F, Pierangeli D, Parravicini J, Claudio Conti AJ, Agranat EDR (2015) Observation of diffraction cancellation for nonparaxial beams in the scale-free-optics regime. Phys Rev A 92(1):013835

22. Di Mei F, Caramazza P, Pierangeli D, Di Domenico G, Ilan H, Agranat AJ, Di Porto P, DelRe E (2016) Intrinsic negative mass from nonlinearity. Phys Rev Lett 116(15):153902
23. Garcia V, Fusil S, Bouzehouane K, Enouz-Vedrenne S, Mathur ND, Barthelemy A, Bibes M (2009) Giant tunnel electroresistance for non-destructive readout of ferroelectric states. Nature 460(7251):81–84
24. Glinchuk MD, Eliseev EA, Morozovska AN (2008) Superparaelectric phase in the ensemble of noninteracting ferroelectric nanoparticles. Phys Rev B 78(13):134107
25. Gumennik A, Kurzweil-Segev Y, Agranat AJ (2011) Electrooptical effects in glass forming liquids of dipolar nano-clusters embedded in a paraelectric environment. Opt Mater Express 1(3):332–343
26. Imai T, Miyazu J, Kobayashi J (2014) Measurement of charge density distributions in kta 1–x nb x o 3 optical beam deflectors. Opt Mater Express 4(5):976–981
27. Jeong I-K, Darling TW, Lee J-K, Proffen T, Heffner RH, Park JS, Hong KS, Dmowski W, Egami T (2005) Direct observation of the formation of polar nanoregions in pb (mg 1/3 nb 2/3) o 3 using neutron pair distribution function analysis. Phys Rev Lett 94(14):147602
28. Khomeriki R, Chotorlishvili L, Tralle I, Berakdar J (2016) Positive-negative birefringence in multiferroic layered metasurfaces. Nano Lett 16(11):7290–7294
29. Kim W-H, Son JY, Shin Y-H, Jang HM (2014) Imprint control of nonvolatile shape memory with asymmetric ferroelectric multilayers. Chem Mat 26(24):6911–6914
30. Kleemann W, Schäfer FJ, Rytz D (1985) Diffuse ferroelectric phase transition and long-range order of dilute k ta 1–x nb x o 3. Phys Rev Lett 54(18):2038
31. Kogelnik H (1969) Coupled wave theory for thick hologram gratings. Bell Labs Tech J 48(9):2909–2947
32. Kounga AB, Granzow T, Aulbach E, Hinterstein M, Rödel J (2008) High-temperature poling of ferroelectrics. J Appl Phys 104(2):024116
33. Kutnjak Z, Petzelt J, Blinc R (2006) The giant electromechanical response in ferroelectric relaxors as a critical phenomenon. Nature 441(7096):956–959
34. Lawes G, Brooks Harris A, Tsuyoshi Kimura N, Rogado RJ, Cava AA, Entin-Wohlman O, Yildirim T, Michel Kenzelmann C, Broholm, et al (2005) Magnetically driven ferroelectric order in ni 3 v 2 o 8. Phys Rev Lett 95(8):087205
35. Lee HN, Christen HM, Chisholm MF, Rouleau CM, Lowndes DH (2005) Strong polarization enhancement in asymmetric three-component ferroelectric superlattices. Nature 433(7024):395–399
36. Li BL, Liu XP, Fang F, Zhu JL, Liu J-M (2006) Monte carlo simulation of ferroelectric domain growth. Phys Rev B 73(1):014107
37. Li H, Tian H, Gong D, Meng Q, Zhou Z (2013) High dielectric tunability of kta0. 60nb0. 40o3 single crystal. J Appl Phys 114(5):054103
38. Li P-F, Tang Y-Y, Wang Z-X, Ye H-Y, You Y-M, Xiong R-G (2016) Anomalously rotary polarization discovered in homochiral organic ferroelectrics. Nat Commun 7:13635
39. Li Q, Cao Y, Yu P, Vasudevan RK, Laanait N, Tselev A, Xue F, Chen LQ, Maksymovych P, Kalinin SV et al (2015) Giant elastic tunability in strained bifeo3 near an electrically induced phase transition. Nat Commun 6
40. Li YL, Hu SY, Liu ZK, Chen LQ (2001) Phase-field model of domain structures in ferroelectric thin films. Appl Phys Lett 78(24):3878–3880
41. Lines ME, Glass AM (1977) Principles and applications of ferroelectrics and related materials. Oxford University Press, Oxford
42. Qieni L, Li B, Li Z, Ge B (2017) Field-induced lifetime enhancement of photorefractive gratings in a mn: Fe: Ktn crystal. Opt Lett 42(13):2407–2410
43. Lummen TTA, Yijia G, Wang J, Lei S, Xue F, Kumar A, Barnes AT, Barnes E, Denev S, Belianinov A et al (2014) Thermotropic phase boundaries in classic ferroelectrics. Nat Commun 5:2014
44. Manley ME, Lynn JW, Abernathy DL, Specht ED, Delaire O, Bishop AR, Sahul R, Budai JD (2014) Phonon localization drives polar nanoregions in a relaxor ferroelectric. Nat Commun 5

45. Meng X, Tian H, Tan P, Huang F, Zhang R, Zhou Z (2017) Strong electromechanical coupling in paraelectric kta1-xnbxo3 crystals. J Am Ceram Soc
46. Menzel C, Helgert C, Rockstuhl C, Kley E-B, Tünnermann A, Pertsch T, Lederer F (2010) Asymmetric transmission of linearly polarized light at optical metamaterials. Phys Rev Lett 104(25):253902
47. Balthasar Mueller JP, Rubin NA, Devlin RC, Groever B, Capasso F (2017) Metasurface polarization optics: Independent phase control of arbitrary orthogonal states of polarization. Phys Rev Lett 118(11):113901
48. Parravicini J, Martínez Lorente R, Di Mei F, Pierangeli D, Agranat AJ, DelRe E (2015) Volume integrated phase modulator based on funnel waveguides for reconfigurable miniaturized optical circuits. Opt Lett 40(7):1386–1389
49. Parravicini J, DelRe E, Agranat AJ, Parravicini G (2016) Macroscopic response and directional disorder dynamics in chemically substituted ferroelectrics. Phys Rev B 93(9):094203
50. Parravicini J, DelRe E, Agranat AJ, Parravicini G (2017) Liquid-solid directional composites and anisotropic dipolar phases of polar nanoregions in disordered perovskite. Nanoscale
51. Phelan D, Stock C, Rodriguez-Rivera JA, Chi S, Leão J, Long X, Xie Y, Bokov AA, Ye Z-G, Ganesh P et al (2014) Role of random electric fields in relaxors. Proc Natl Acad Sci 111(5):1754–1759
52. Pierangeli D, Di Mei F, Parravicini J, Parravicini GB, Agranat AJ, Conti C, DelRe E (2014) Observation of an intrinsic nonlinearity in the electro-optic response of freezing relaxors ferroelectrics. Opt Mater Express 4(8):1487–1493
53. Pierangeli D, Di Mei F, Conti C, Agranat AJ, DelRe E (2015) Spatial rogue waves in photorefractive ferroelectrics. Phys Rev Lett 115(9):093901
54. Pierangeli D, Flammini M, Di Mei F, Parravicini J, de Oliveira CEM, Agranat AJ, DelRe E (2015) Continuous solitons in a lattice nonlinearity. Phys Rev Lett 114(20):203901
55. Pierangeli D, Di Mei F, Di Domenico G, Agranat AJ, Conti C, DelRe E (2016a) Turbulent transitions in optical wave propagation. Phys Rev Lett 117(18):183902
56. Pierangeli D, Ferraro M, Di Mei F, Di Domenico G, de Oliveira CEM, Agranat AJ, DelRe E (2016) Spontaneous photonic super-crystals in composite ferroelectrics. In: 2016 conference on lasers and electro-optics (CLEO). IEEE, pp 1–2
57. Pirc R, Kutnjak Z (2014) Electric-field dependent freezing in relaxor ferroelectrics. Phys Rev B 89(18):184110
58. Pirc R, Blinc R (2007) Vogel-fulcher freezing in relaxor ferroelectrics. Phys Rev B 76(2):020101
59. Plum E, Liu X-X, Fedotov VA, Chen Y, Tsai DP, Zheludev NI (2009) Metamaterials: optical activity without chirality. Phys Rev Lett 102(11):113902
60. Potter BG Jr, Tikare V, Tuttle BA (2000) Monte carlo simulation of ferroelectric domain structure and applied field response in two dimensions. J Appl Phys 87(9):4415–4424
61. Prosandeev S, Dawei Wang AR, Akbarzadeh BD, Bellaiche L (2013) Field-induced percolation of polar nanoregions in relaxor ferroelectrics. Phys Rev Lett 110(20):207601
62. Pugachev AM, Kovalevskii VI, Surovtsev NV, Kojima S, Prosandeev SA, Raevski IP, Raevskaya SI (2012) Broken local symmetry in paraelectric batio 3 proved by second harmonic generation. Phys Rev Lett 108(24):247601
63. Rabe KM, Ahn CH, Triscone J-M (2007) Physics of ferroelectrics: a modern perspective, vol 105. Springer Science & Business Media, Berlin
64. Ramachandran GN (1964) Advanced methods of crystallography. Academic Press Inc, New York
65. Ren M-X, Wei W, Cai W, Pi B, Zhang X-Z, Jing-Jun X (2017) Reconfigurable metasurfaces that enable light polarization control by light. Light Sci Appl 6(6)
66. Roppo V, Wang W, Kalinowski K, Kong Y, Cojocaru C, Trull J, Vilaseca R, Scalora M, Krolikowski W, Kivshar Y (2010) The role of ferroelectric domain structure in second harmonic generation in random quadratic media. Opt Express 18(5):4012–4022
67. Sakamoto T, Sasaura M, Yagi S, Fujiura K, Cho Y (2008) In-plane distribution of phase transition temperature of kta1-xnbxo3 measured with single temperature sweep. Appl Phys Express 1(10):101601

68. Shevchenko A, Roussey M, Friberg AT, Setälä T (2017) Polarization time of unpolarized light. Optica 4(1):64–70
69. Shvartsman VV, Lupascu DC (2012) Lead-free relaxor ferroelectrics. J Am Ceram Soc 95(1):1–26
70. Takagi M, Ishidate T (2000) Anomalous birefringence of cubic batio 3. Solid State Commun 113(7):423–426
71. Tan P, Tian H, Chengpeng H, Meng X, Mao C, Huang F, Shi G, Zhou Z (2016) Temperature field driven polar nanoregions in kta1- x nb x o3. Appl Phys Lett 109(25):252904
72. Tan P, Tian H, Mao C, Chengpeng H, Meng X, Li L, Shi G, Zhou Z (2017) Field-driven electro-optic dynamics of polar nanoregions in nanodisordered kta1- x nb x o3 crystal. Appl Phys Lett 111(1):012903
73. Tian H, Yao B, Hu C, Meng X, Zhou Z (2014) Impact of polar nanoregions on the quadratic electro-optic effect in k0. 95na0. 05ta1- xnbxo3 crystals near the curie temperature. Appl Phys Express 7(6):062601
74. Tian H, Tan P, Meng X, Chengpeng H, Yao B, Shi G, Zhou Z (2015) Variable gradient refractive index engineering: design, growth and electro-deflective application of kta 1–x nb x o 3. J Mater Chem C 3(42):10968–10973
75. Tian H, Yao B, Tan P, Zhou Z, Shi G, Gong D, Zhang R (2015) Double-loop hysteresis in tetragonal kta0. 58nb0. 42o3 correlated to recoverable reorientations of the asymmetric polar domains. Appl Phys Lett 106(10):102903
76. Tian H, Yao B, Wang L, Tan P, Meng X, Shi G, Zhou Z (2015) Dynamic response of polar nanoregions under an electric field in a paraelectric kta0. 61nb0. 39o3 single crystal near the para-ferroelectric phase boundary. Sci Rep 5:13751
77. Toulouse J, DiAntonio P, Vugmeister BE, Wang XM, Knauss LA (1992) Precursor effects and ferroelectric macroregions in kta 1–x nb x o 3 and k 1- y li y tao 3. Phys Rev Lett 68(2):232
78. Trull J, Cojocaru C, Fischer R, Saltiel SM, Staliunas K, Herrero R, Vilaseca R, Neshev DN, Krolikowski W, Kivshar YS (2007) Second-harmonic parametric scattering in ferroelectric crystals with disordered nonlinear domain structures. Opt Express 15(24):15868–15877
79. Wang L, Tian H, Meng X, Chen H, Zhou Z, Shen Y (2014) Field-induced enhancement of voltage-controlled diffractive properties in paraelectric iron and manganese co-doped potassium-tantalate-niobate crystal. Appl Phys Express 7(11):112601
80. Weber MF, Stover CA, Gilbert LR, Nevitt TJ, Ouderkirk AJ (2000) Giant birefringent optics in multilayer polymer mirrors. Science 287(5462):2451–2456
81. Wemple SH, DiDomenico M Jr (1969) Oxygen-octahedra ferroelectrics. ii. Electro-optical and nonlinear-optical device applications. J Appl Phys 40(2):735–752
82. Westphal V, Kleemann W, Glinchuk MD (1992) Diffuse phase transitions and random-field-induced domain states of the "relaxor" ferroelectric pbmg 1/3 nb 2/3 o 3. Phys Rev Lett 68(6):847
83. Guangyong X, Zhong Z, Bing Y, Ye Z-G, Shirane G (2006) Electric-field-induced redistribution of polar nano-regions in a relaxor ferroelectric. Nat Mater 5(2):134–140
84. Yariv A, Yeh P (1984) Optical waves in crystals, vol 10. Wiley, New York
85. Zhu W, Chao J-H, Chen C-J, Yin S, Hoffman RC (2016) Three order increase in scanning speed of space charge-controlled KTN deflector by eliminating electric field induced phase transition in nanodisordered KTN. Sci Rep 6

Chapter 8
Intrinsic Negative-Mass from Nonlinearity

Scale-free-optics, or diffraction-cancellation is a propagation regime (discovered by *DelRe* et al. in [1], but first observation can be traced back to [2]) in which the electromagnetic fields is no longer governed by the Helmholtz equation but by a Klein–Gordon-type equation. This is achieved in a disordered out of equilibrium para-electric crystal near the phase transition. In this condition beam propagation is affected by a giant and purely diffusive nonlinearity which has profound implications for wave dynamics. In particular in this regime optical propagation occurs without any limit associated to the optical wavelength [3] (scale-free-optics), where the diffraction is absent, not simply compensated by nonlinear index change or the presence of waveguide (both conditions in which the spatial dimensions scale with λ). The phenomenon appears also to be intensity and size independent [4], but it is nonetheless nonlinear. Experiments that highlight the nonlinear nature of diffraction cancellation are beam-beam interaction phenomena, which involve beam attraction, crossing, and beam spiraling, three interaction phenomena that are similar to those normally associated to solitons [5]. The unique features of the system allow us to observe anti-diffraction of light and light beams that can be focused to dimensions smaller than the diffraction limit [6, 7].

In this chapter we propose and provide experimental evidence of a mechanism able to support negative intrinsic effective mass. The idea is to use anti-diffraction to change the sign of the mass in the leading linear propagation equation. Intrinsic negative mass dynamics is reported for light beams in a ferroelectric crystal substrate, where the diffusive photorefractive nonlinearity leads to a negative-mass Schrodinger Equation. The signature of inverted dynamics is the observation of beams repelled from strongly guiding integrated waveguides irrespective of wavelength and intensity and suggests shape-sensitive nonlinearity as a basic mechanism leading to intrinsic negative mass. This work has been published in the article published in *'Intrinsic negative mass from nonlinearity'* Physical Review Letters **116**, 153902 (2016).

G. Di Domenico, *Electro-optic Photonic Circuits*, Springer Theses,
https://doi.org/10.1007/978-3-030-23189-7_8

8.1 Introduction

A negative energy density is thought to play a key role in cosmological conjectures, such as in stabilizing space-time wormholes and in explaining the supposed acceleration of the expanding universe. At present, there is no proposed mechanism to support negative mass as a local space-time property. We here describe a mechanism able to support intrinsic negative mass through nonlinearity and provide experimental evidence of inverted dynamics for a light beam in a nanodisordered lithium-enriched potassium-tantalate-niobate (KTN:Li) crystal waveguide. The effect does not involve periodicity and, being intrinsic to the beam, is not limited to specific directions or energies.

Negative mass particles should be repelled from attractive potentials and attracted from repelling ones, an unfathomed physics that could revolutionize our picture of nature, rendering abstract conjectures, such as space-time wormholes, at least in principle stable [8, 9]. With a mass $m < 0$, the particle subject to a potential U suffers a force $F = -\nabla U$ but manifests the inverted acceleration $a = \nabla U/|m|$ (Fig. 8.1). Although all known particles have a positive or zero mass, conditions can be found in which the interaction of a particle with its environment leads to an effective mass $m^* \neq m$ that can, in precise conditions, also be negative. To date, $m^* < 0$ has been demonstrated in periodic systems [10–14], where the periodicity in the $\epsilon(k)$ band structure causes there to be a finite region of wave-vectors for which the Bloch-modes have a constant negative $d^2\epsilon/dk^2 < 0$ and with it, a behavior described by a negative effective mass $m^* = \hbar^2(d^2\epsilon/dk^2)^{-1} < 0$ [15]. Intuitively, internal components move out-of-phase with respect to the global resonance of the system and lead to a negative momentum response for a positive-momentum excitation [16]. Negative mass in these periodic systems is not intrinsic to the particle or wave, but only occurs for precise wavevectors at the edge of the Brillouin zone. At present, no mechanism has been proposed and demonstrated able to support negative mass as a property of a localized wave with inverted dynamics irrespective of particle energy or wave-vector.

8.2 Negative Mass from Theory

Consider the Schroedinger Equation (SE)

$$(i\partial_t + (\hbar/2m)\nabla^2)\psi = (V/\hbar)\psi, \tag{8.1}$$

where $m > 0$. As an axiom, the SE is linear, but assume that there is some mechanism that violates this linearity so that, in general, the potential has two components, $V + V_{nl}$, with V just a standard potential and V_{nl} a specific form of self-action. Indeed, although nonlinearity is absent in quantum mechanics, it is naturally built into the

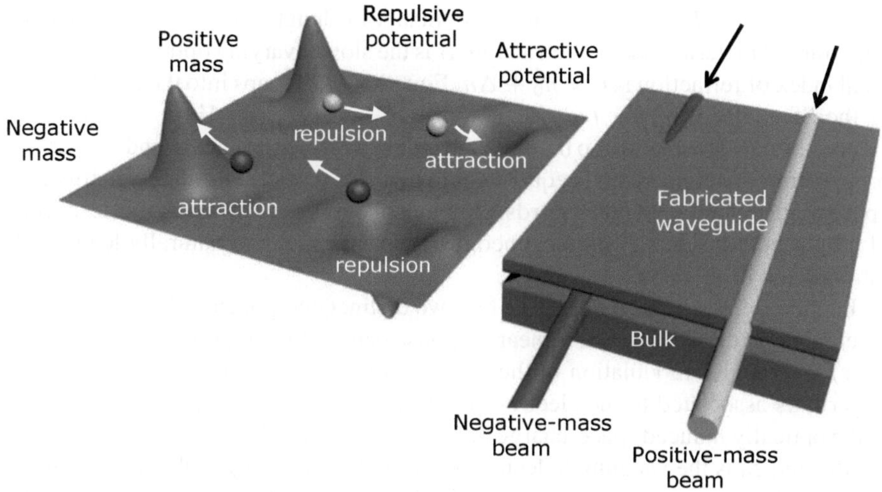

Fig. 8.1 (Color online) Intrinsic inverted dynamics and a negative-mass Schroedinger Equation (SE). (Left) In constrast to a positive mass particle (lightly-shaded sphere), a negative mass particle (dark sphere) will be attracted by a repelling potential and repelled by a binding one. (Right) A propagating light beam described by a positive-mass SE (lightly-shaded beam) will be guided by an integrated waveguide whereas a negative-mass light beam (dark beam) will be repelled by it and scattered into the bulk of the substrate

Einstein equations for which negative mass may have some important role. If V_{nl} is a small local perturbation associated to $|\psi|^2$, such as a Kerr effect with $V_{nl} \propto |\psi|^2$, the SE turns into a Nonlinear-Schroedinger Equation that supports solitons and rogue waves, but for which no negative mass dynamics emerges. Nonlinearity, in turn, can take many forms and also be nonlocal, involving integrals and derivatives of $|\psi|^2$. If self-action is approximated by $V_{nl}\psi \simeq (\hbar/2m')\nabla^2\psi$, then this will radically transform Eq. (8.1) into

$$(i\partial_t + (\hbar/2m^*)\nabla^2)\psi = (V/\hbar)\psi \tag{8.2}$$

that, when $m' < m$, corresponds to a negative mass SE with $m^* = mm'/(m' - m) < 0$.

In our experiments we observe inverted dynamics in the propagation of light in a nanodisordered ferroelectric with the combined effect of an integrated slab waveguide and diffusive photorefractive nonlinearity [1, 17, 18]. Paraxial propagation along the z axis is governed by the parabolic equation

$$(i\partial_z + (1/2k)\nabla_\perp^2)A = -(k\Delta n/n_0)A, \tag{8.3}$$

where $k = k_0 n_0$, $k_0 = 2\pi/\lambda$, λ is the optical wavelength ($\omega = 2\pi c/\lambda$), n_0 is the unperturbed material index of refraction, A is the slowly varying optical field, and the local index of refraction is $n = n_0 + \Delta n$. Equation (8.3) maps into the 2+1D version of the SE of Eq. (8.2) for $t \equiv z/c$, $\psi \equiv A$, $m^* \equiv \hbar k/c$, and $V \equiv -\hbar c(k\Delta n/n_0)$. Hence, the causal relationship between the index of refraction pattern and the paraxial propagation of a light beam is equivalent to that of a particle with finite energy in an appropriate potential. In other words, although photons have no mass, the description of a light field inside an inhomogeneous transparent material naturally leads to the introduction of $m^* \neq 0$.

In our case, the index modulation has two distinct components, $\Delta n = \delta n_{nl} + \delta n$, where δn_{nl} is the diffusive nonlinear response caused by the propagating light, and $\delta n(\mathbf{r})$ is the index modulation of the fabricated slab waveguide. The diffusive nonlinearity is associated to the electro-optic response $\delta n_{nl} = -(n_0^3/2)g\epsilon_0^2\chi_{PNR}^2|\mathbf{E}_{dc}|^2$ to the optically induced space-charge field $\mathbf{E}_{dc}$, where g is the quadratic electro-optic coefficient, ε_0 is the vacuum dielectric permittivity, and χ_{PNR} is the low-frequency susceptibility of the polar-nanoregions (PNRs) [19, 20]. With no external bias, photoexcited charge diffusion causes $\mathbf{E}_{dc} = -(k_B T/q)\nabla I/I$, where k_B is the Boltzmann constant, T the crystal temperature, q the elementary charge, and $I = |A|^2$ is the intensity of the optical field A. The propagation equation now reads [1, 18]

$$\left(i\partial_z + \frac{1}{2k}\nabla_\perp^2\right)A = -\frac{k\delta n}{n_0}A + \frac{1}{2k}\frac{L^2}{4\lambda^2}\left(\frac{\nabla_\perp I}{I}\right)^2 A, \tag{8.4}$$

the Helmholtz Equation $\nabla^2 E + k_0^2 n_0^2 E + 2k_0^2 n_0 \delta n E + 2k_0^2 n_0 \Delta n_{nl} E = 0$ reads

$$\nabla^2 E + k_0^2 n_0^2 E + 2k_0^2 n_0 \delta n E - \frac{L^2}{4\lambda^2}\left(\frac{\nabla I}{I}\right)^2 E = 0, \tag{8.5}$$

where $L = 4\pi n_0^2 \epsilon_0 \sqrt{g}\chi_{PNR}(k_B T/q)$, that, for Gaussian-like beams, is well approximated by the linear wave equation

$$(i\partial_z + (1/2k)(1 - L^2/\lambda^2)\nabla_\perp^2)A = -(k\delta n/n_0)A. \tag{8.6}$$

This propagation equation, rewritten as

$$(i\partial_z - (1/2\tilde{k})\nabla_\perp^2)A = -(k/n_0)\delta n A, \tag{8.7}$$

with

$$\tilde{k} = \frac{k}{\frac{L^2}{\lambda^2} - 1}, \tag{8.8}$$

for $L > \lambda$ Eq. (8.6) maps to the 2+1D version of the SE of Eq. (8.2) with

$$(i\partial_t + (\hbar/2m)\nabla_\perp^2)\psi = (V/\hbar)\psi, \tag{8.9}$$

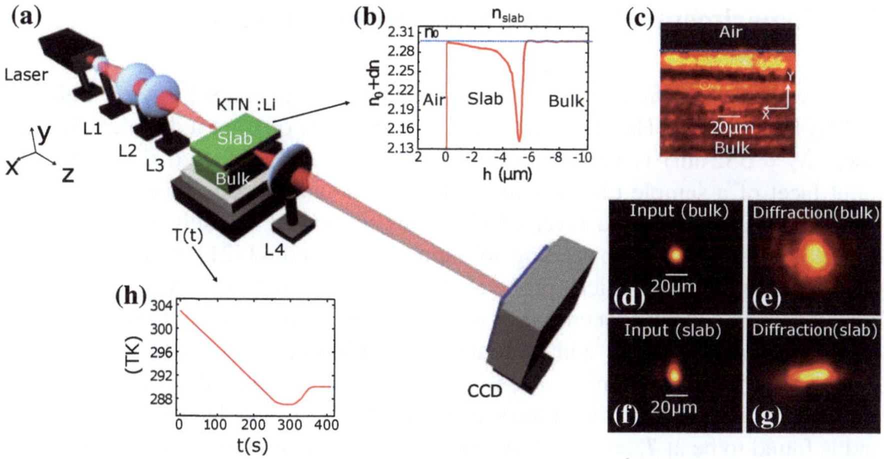

Fig. 8.2 Experimental setup, materials and protocol. **a** A laser beam is launched into the KTN:Li waveguide and imaged on a CCD using lenses L1-L4 (L4 has NA$\simeq$ 0.35). **b** Waveguide index profile (for 532 nm). **c** Typical graded waveguide output intensity distribution for an expanded plane-wave input. **d–g** Input and diffraction intensity distribution pattern in the bulk crystal and in the waveguide ($L/\lambda \simeq 0$). At a constant $T_A = 303$ K, the 100 μW input beam (waist $w_{0x} \sim w_{0y} = 8\,\mu$m) **d** diffracts to 22 μm **e** after propagating a distance of $L_z \simeq 2.4$ mm through the bulk crystal. In the waveguide, the input beam **f** diffracts to $w_{0x} = 9.8$ and $w_{0y} = 31.7\,\mu$m (**g**). **h** Supercooling protocol $T(t)$ to achieve $L/\lambda > 1$

for $t \equiv z/c$, $\psi \equiv A$, $m \equiv -\hbar\tilde{k}/c < 0$, $V(x, y) \equiv -\delta n(x, y)k\hbar c/n_0$, and

$$m^* = -\frac{\hbar k}{c}\frac{1}{\frac{L^2}{\lambda^2} - 1} < 0. \tag{8.10}$$

The $m^* < 0$ regime is here a product of nonlinearity, is localized around the beam, and is not limited to specific wavelengths, directions, or resonances of the system. We note that the passage from the nonlinear Eq. (8.4) to the linear Eq. (8.6) is rigorously valid only for Gaussian beams for which the peak intensity factors out of the term $(\nabla_\perp I/I)$. Consistently, even though beams may be spreading or becoming tighter during propagation, they will have only one specific value of m^* (as per Eq. (8.10)). Since the passage to Eq. (8.6) is valid for Gaussian beam shapes, it follows that the effective negative mass will arise only if the δn is comparable or larger to the Gaussian beam itself. Fabricated waveguides considerably smaller than the beam waist will correspond to a potential well as in Fig. 8.1 that is smaller than the size of the particle itself and not necessarily lead to inverted dynamics. A flag to this spatial requirement is that Eq. (8.4) is spatially nonlocal whereas Eq. (8.6) is not.

8.3 Experiments: Negative Mass in Slab Waveguide

We carry out experiments with the setup illustrated in Fig. 8.2a. An x-polarized TEM_{00} beam from a He-Ne laser with ($\lambda_1 = 633$ nm) or from a doubled Nd:YAG laser ($\lambda_2 = 532$ nm) is first expanded and subsequently focused down onto the input facet of a sample of potassium-lithium-tantalate-niobate doped with copper (KLTN:Cu) crystal with a layer of He^+ ions implanted beneath its surface. The crystal is grown by the top seeding solution growth method [21]. Its composition is determined by electron micro-probe analysis and is found to be $K_{0.985}Li_{0.015}Ta_{0.63}Nb_{0.37}O_3$. The copper concentration is determined by Inductively Coupled Plasma (ICP) mass spectrometry and is found to be 68 ppm (in weight). A sample of $3.9^{(x)} \times 0.9^{(y)} \times 2.4^{(z)}\,\mathrm{mm}^3$ in size is cut along the [001] crystallographic axis. The ferroelectric phase transition of the sample is derived from dielectric measurements, and is found to be at $T_c = 285$ K. At the operating temperature range of 286–305 K the sample maintains high optical quality with refractive index of $n_0 = 2.3$, and quadratic electro optic coefficient of $0.14\,\mathrm{m}^4\mathrm{C}^{-2}$. The He^+ ions are implanted at 2.3 MeV with fluence of $0.8 \cdot 10^{16}$ ions/cm^2 which yields a partially amorphous layer with refractive index distribution as presented in Fig. 8.2b [22, 23]. This forms a slab waveguide between the surface of the sample and the implanted layer that acts as the cladding [22]. The transverse intensity distribution of the beam is imaged using a CCD camera through the imaging lens. The diffraction pattern at the output facet of the crystal $L/\lambda \simeq 0$, in the bulk and in the slab waveguide respectively, is shown in Fig. 8.2d–g. In Fig. 8.2h we report the thermal shock protocol $T(t)$ near the peak in the dielectric response at $T_m = 287.5$ K [4, 24–26] that allows a transient $L/\lambda > 1$. In practice, the crystal is first cleaned of photorefractive space-charge by illuminating it with a microscope illuminator. Using a temperature controller that drives the current of a Peltier junction placed directly below the crystal in the y direction, we bring the sample to thermalize at $T_A = 303$ K. The sample is then cooled at the rate of 0.06 K/s to a temperature $T_D = 287$ K, where it is kept for 60 s. It is then reheated at a rate of 0.1 K/s to the operating temperature ($>T_D$) $T_B = 290$ K. The Peltier junction is placed below the sample so that during the process the crystal, exposed to ambient air (at ~290 K), experiences a transient temperature gradient along the y axis. Once T_B is reached and the temperature cycle $T(t)$ is complete, we switch on the laser beam, recording front view images of the intensity distribution.

In Fig. 8.3a we report the basic signature of intrinsic negative mass SE dynamics: a beam expelled from the fabricated waveguide and scattered into the substrate. The beam is launched into the waveguide at $t = 0$ after the sample has undergone supercooling (the $T(t)$ in Fig. 8.2h). It is first observed to focus down, anti-diffract, and then suffer a strong repulsion, when it is scattered into the metastable bulk. Ultimately, the beam is observed to relax back into a linear diffraction, diffracting in the x direction and guided in the y. The sequence of events is further detailed in Fig. 8.3b, where the beam peak intensity is plotted versus time. For comparison, we include the same curve when the same beam is launched into the bulk of the substrate. In the slab the beam suffers a transient scattering, whereas in the bulk it suffers anti-

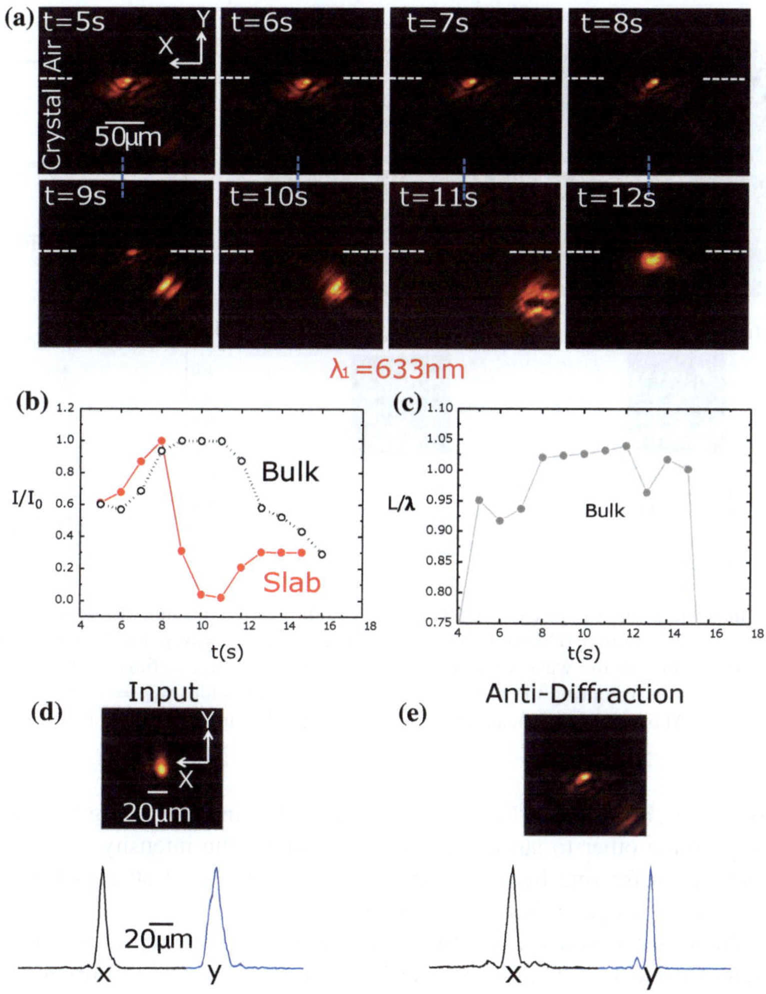

Fig. 8.3 A fabricated waveguide repels light as it acquires an intrinsic negative mass. **a** Time sequence of the output intensity distributions for $\lambda_1 = 633$ nm. **b** Comparison between the maximum peak intensity of the beam in the waveguide and in the bulk during the transient. **c** Time dependence of the L/λ in bulk. **d–e** Transient anti-diffraction in the waveguide: the input beam (waist $w_{0x} = 9.9$, $w_{0y} = 9.3\,\mu$m) (**d**) and the output beam (minimum $w_{0x} = 6.8$, $w_{0y} = 7.1\,\mu$m, $L/\lambda \simeq 1.05$) during the after-shock (**e**)

diffraction dynamics [6, 7]. The connection between this transient repulsion from the waveguide and the change in sign of the beam mass in the equivalent SE is investigated in Fig. 8.3c. Using the bulk anti-diffraction and the analytical anti-diffraction theory, L/λ as a function of time is evaluated. As expected, the instants of time during which the dipolar relaxation leads to $L/\lambda > 1$ coincide with the repulsive regime. In other words, the behavior of the light beam is drastically different between the guided and

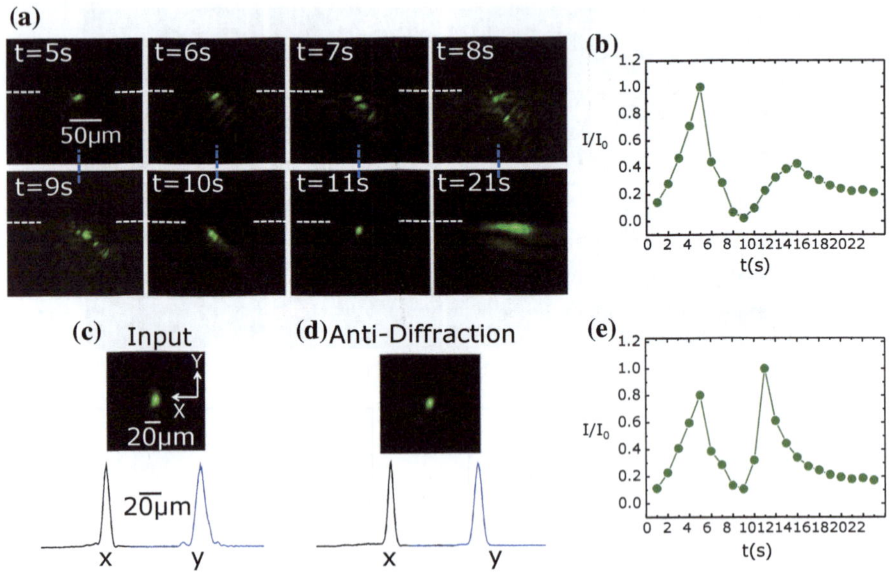

Fig. 8.4 Intrinsic negative mass dynamics in the waveguide for $\lambda_2 = 532$ nm. **a** Time sequence of the output intensity distribution. **b** Maximum beam peak intensity in the waveguide during the transient. **c–d** Anti-diffraction in the waveguide: **c** the input beam (waist $w_{0x} = 6.8\,\mu$m ,$w_{0y} = 9.9\,\mu$m) and **d** the output beam during the after-shock (minimum beam width $w_{0x} = 7.1\,w_{0y} = 5.6\,\mu$m, $L/\lambda \simeq 1.04$). **e** Maximum beam peak intensity of the output beam during the transient (see text)

bulk conditions, as shown in Fig. 8.3b, in one case leading to a strong repulsion and scattering, in the other to strong spatial localization. The intensity distribution of input and anti-diffracting light corresponding to the $L/\lambda > 1$ stage before light is repelled by the waveguide is reported in Fig. 8.3d–e.

To validate the negative mass SE model of Eq. (8.6) we repeated experiments for different intensities. The strong transient response reported in Figs. 8.3 and 8.4 has a characteristic response time of tens of seconds. Experiments using beams with different powers (10, 20, 40, 80, 100 μW) lead to similar results and time scales. This approximate intensity-independent nature of the phenomenon is compatible with the overall linear nature of the effect as described in Eq. (8.2). Weak dependence of time scales on peak intensity indicates that time dynamics are principally associated to the relaxation of the metastable PNR state, while the photorefractive build-up is relatively faster and the space-charge field can be considered at all times at steady state, corroborating the validity of the diffusive nonlinearity model. The value of the L parameter is estimated by measuring the output and input waist ratio.

To prove the effect is not limited to a specific region of wavevectors, in Fig. 8.4 we report beam repulsion for $\lambda = 532$ nm. The effect is analogous to the previous one, even though the details of the time evolution vary for each thermal shock, and only an average relaxation has a precise dynamical meaning. Specifically, the estimated value of L/λ in the two cases of Figs. 8.3 and 8.4 is comparable even though the thermal shock is the same and the wavelengths are different. Fluctuations are further exalted during the transition from the diffractive positive mass SE to the anti-diffractive negative mass SE, as the waveguide goes from being guiding to anti-guiding and allows light to explore its surroundings. An interesting difference in the dynamics of Figs. 8.3 and 8.4 is that the shorter wavelength case manifests a second focused stage reported in Fig. 8.4e, displaced outside the original waveguide, where no second peak is found (Fig. 8.4b). Precisely, the second peak is displaced approximately 4 μm in the y direction, inside the amorphous region (see Fig. 8.2b). This indicates that the antiguiding amorphous layer becomes guiding in the negative effective mass regime. Unfortunately, the amorphous layer is only $\simeq$1 μm wide and its effect on the beam cannot be fully described by the passage from Eqs. (8.4) to (8.6). Congruently, for the longer wavelength cases of Fig. 8.3, no analogous effect is observed.

Numerical simulations of the stationary full-nonlinear Eq. (8.4) are performed with a split-step Fourier method and with parameters matching our experimental conditions and slab-waveguide profile; results agree well with our observations and are reported in Fig. 8.5. They allow us to inspect the details of the propagation during evolution (Fig. 8.5a–d) that cannot be directly detected optically and the resilience of the effect on distortions in the input Gaussian beam shape (Fig. 8.5e). In particular, the transition from positive to negative-mass dynamics is well reproduced as a function of (L/λ), with the expulsion of the beam from the waveguide to the substrate for $(L/\lambda) > 1$. We note that this expulsion is fundamentally different respect to the phenomenon of soliton ejection and tunneling from a potential where the refractive-index well is modified by the nonlinear dynamics [27–29]. In our present phenomenon, no available nonlinearity could even marginally modify the huge fabricated index modulation (index modulations up to $\delta n \simeq 0.15$), and expulsion is a consequence of a change in the sign of the effective-mass of the light beam.

8.4 Conclusion

We have discussed how a shape-sensitive nonlinearity can lead to an intrinsic negative mass localized around a wave without the constraints associated to a periodic system or a resonance. We have shown an instance in which light-matter interaction, instead of modifying the nature of the propagation equation introducing nonlinear terms that alter wave-propagation and lead to solitons, shock-waves and rogue-waves [30–34], causes light to obey a modified linear propagation equation of the type $(i\partial_z - (1/2\tilde{k})\nabla_\perp^2)A = 0$, with $\tilde{k} > 0$. The effect hinges on a transient anti-diffraction that is a product of nonlinearity and requires no underlying periodic pattern, in distinction to linear anti-diffraction [35–38]. In these conditions, beam propagation naturally maps the dynamics of a negative mass particle described by the Schroedinger Equation with $m^* = -\hbar\tilde{k}/c < 0$, so that a fabricated waveguide with a strong guiding index modulation $\delta n/n \sim 10\%$, amounting to a strong binding potential, repels light instead of attracting it. In distinction to previous studies into negative mass, that focus on effective dynamics in periodic potentials, our study here

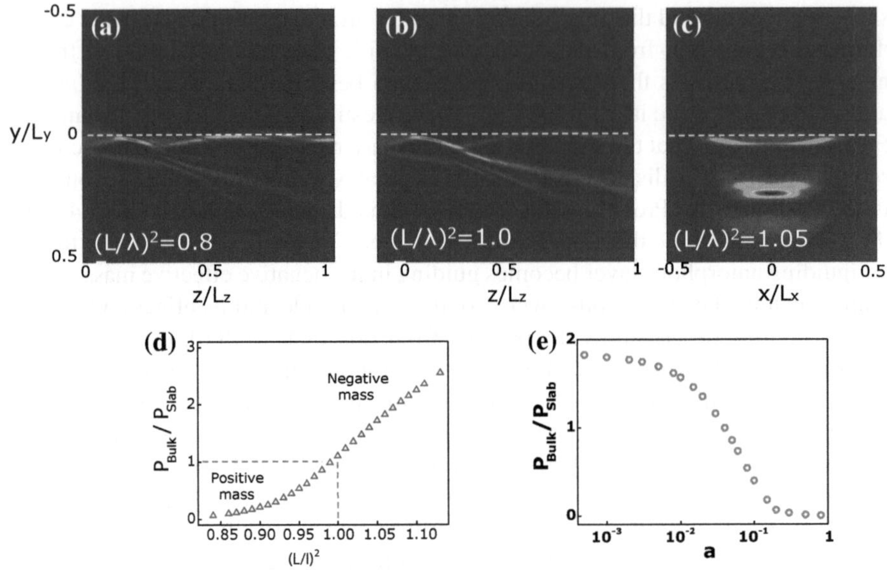

Fig. 8.5 Numerical simulation of Eq. (8.4). Dynamics of the beam along z ($L_x = 80\,\mu$m, L_z =2.4 mm, $\lambda = 633$ nm) **a** for $(L/\lambda)^2 = 0.85$ and **b** $(L/\lambda)^2 = 1.05$. **c** Output intensity distribution in the negative effective mass case. **d** Ratio of total power scattered into the bulk P_{Bulk} to that retained by the slab P_{Slab} versus $(L/\lambda)^2$. An effective positive mass is compatible with $P_{Bulk}/P_{Slab} < 1$, whereas a negative mass is compatible with inverted dynamics and $P_{Bulk}/P_{Slab} > 1$. **e** Dependence of inverted dynamics on beam shape: the P_{Bulk}/P_{Slab} ratio for ever more distorted and squared-off Gaussian inputs ($\exp[-(x^2 + y^2)/w_0^2) - a(x^4 + y^4)/w_0^4]$, w_0 is the input beam width)

demonstrates a local mechanism that provides negative mass compatible with basic negative mass conjectures, such as those required to stabilize space-time wormholes.

8.5 Supplementary Information

We here report supporting experiments on the role of the fabricated waveguide in the repulsion and scattering of the anti-diffracting light beam.

Anti-diffraction of y-Polarized Beams

In Fig. 8.6 we report the output intensity distribution in the same experimental conditions of Fig. 8.3, with now a y polarized beam. The guiding properties of the fabricated waveguide are weaker and congruently no negative-mass regime with a strong repulsion of light into the substrate is observed following the anti-diffraction stage.

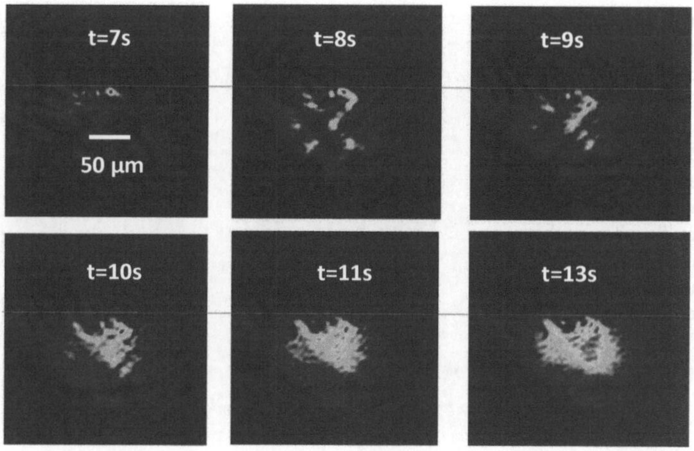

Fig. 8.6 Output intensity distribution for a y-polarized beams gives rise to anti-diffraction without the negative-mass stage with the signature strong scattering into the substrate. False colors are used to highlight less intense regions of the beam

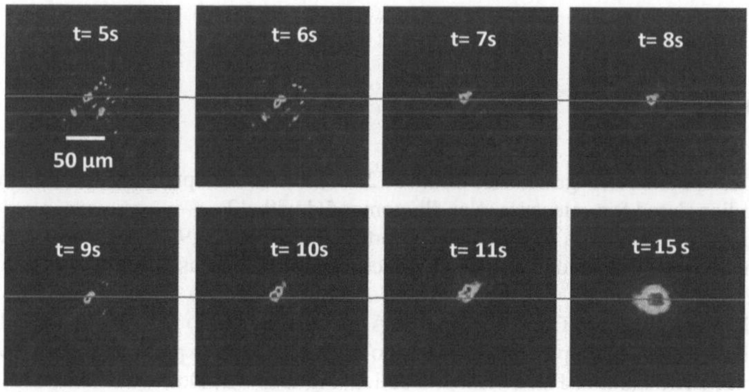

Fig. 8.7 Output intensity distribution indicates an absence of negative-mass dynamics in the bulk of the substrate. False colors are used to highlight less intense regions of the beam

Anti-diffraction in Bulk

In Fig. 8.7 we report the same experiment reported in Fig. 8.3 repeated in the bulk of the substrate (50 μm below the top surface of the sample). As expected, the sequence in time of the dipolar relaxation does not contain the signature scattering away from the original propagation axis. Inverted dynamics are only observed with an external potential in the form of the fabricated pattern.

Critical Scattering During the Phase-Transition

In order to compare results reported in the main text with standard scattering off clusters during the ferroelectric-paraelectric phase-transition, we report in Fig. 8.8

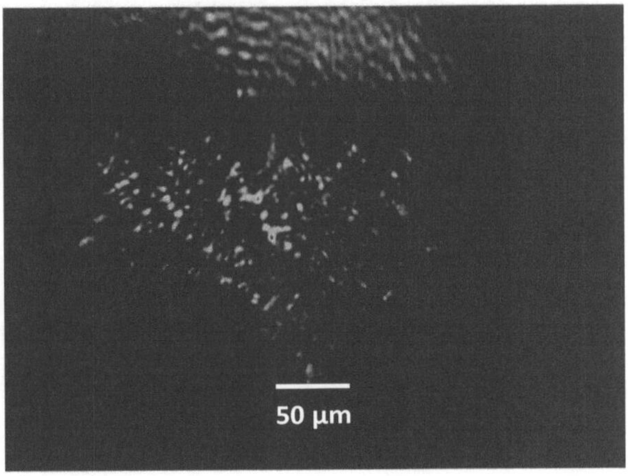

Fig. 8.8 Critical scattering of light during the phase-transition. Light launched at the input inside the waveguide is strongly scattered during propagation when the sample forms ferroelectric clusters at the critical point $T = T_m$, so that the output intensity distribution leads to a speckle-like disordered pattern

transmission images when the crystal is brought to $T = T_m$. Dipolar metastability introduces strong scattering and noise in light propagation associated to local fluctuations of the index of refraction.

References

1. DelRe E, Spinozzi E, Agranat AJ, Conti C (2011) Scale-free optics and diffractionless waves in nanodisordered ferroelectrics. Nat. Photonics 5(1):39–42
2. Crosignani B, Degasperis A, DelRe E, Di Porto P, Agranat AJ (1999) Nonlinear optical diffraction effects and solitons due to anisotropic charge-diffusion-based self-interaction. Phys Rev Lett 82(8):1664
3. Parravicini J, Di Mei F, Conti C, Agranat AJ, DelRe E (2011) Diffraction cancellation over multiple wavelengths in photorefractive dipolar glasses. Opt Express 19(24):24109–24114
4. Di Mei F, Pierangeli D, Parravicini J, Conti C, Agranat AJ, DelRe E (2015) Observation of diffraction cancellation for nonparaxial beams in the scale-free-optics regime. Phys Rev A 92(1):013835
5. Chen Z, Morandotti R (2012) Nonlinear photonics and novel optical phenomena, vol 170. Springer, Berlin
6. DelRe E, Di Mei F, Parravicini J, Parravicini G, Agranat AJ, Conti C (2015) Subwavelength anti-diffracting beams propagating over more than 1,000 rayleigh lengths. Nat Photon
7. Di Mei F, Parravicini J, Pierangeli D, Conti C, Agranat A, DelRe E (2014) Anti-diffracting beams through the diffusive optical nonlinearity. Opt Express 22(25):31434–31439
8. Morris MS, Thorne KS (1988) Wormholes in spacetime and their use for interstellar travel: a tool for teaching general relativity. Am J Phys 56(5):395–412
9. Morris MS, Thorne KS, Yurtsever U (1988) Wormholes, time machines, and the weak energy condition. Phys Rev Lett 61(13):1446
10. Batz S, Peschel U (2013) Diametrically driven self-accelerating pulses in a photonic crystal fiber. Phys Rev Lett 110(19):193901
11. Firstenberg O, Peyronel T, Liang Q-Y, Gorshkov AV, Lukin MD, Vuletić V (2013) Attractive photons in a quantum nonlinear medium. Nature 502(7469):71–75
12. Zhengyou L, Xixiang Z, Yiwei M, Zhu YY, Yang Z, Ting Chan C, Sheng P (2000) Locally resonant sonic materials. Science 289(5485):1734–1736
13. Sakaguchi H, Malomed BA (2004) Dynamics of positive-and negative-mass solitons in optical lattices and inverted traps. J Phys B 37(7):1443

14. Yao Shanshan, Zhou Xiaoming, Gengkai Hu (2008) Experimental study on negative effective mass in a 1d mass-spring system. New J Phys 10(4):043020
15. Charles K (2005) Introduction to solid state physics. Wiley, Amsterdam
16. Wimmer M, Regensburger A, Bersch C, Miri M-A, Batz S, Onishchukov G, Christodoulides DN, Peschel U (2013) Optical diametric drive acceleration through action-reaction symmetry breaking. Nat Phys 9(12):780–784
17. Christodoulides DN, Coskun TH (1996) Diffraction-free planar beams in unbiased photorefractive media. Opt. Lett. 21(18):1460–1462
18. Crosignani B, DelRe E, Di Porto P, Degasperis A (1998) Self-focusing and self-trapping in unbiased centrosymmetric photorefractive media. Opt Lett 23(12):912–914
19. Bokov AA, Ye Z-G (2006) Recent progress in relaxor ferroelectrics with perovskite structure. In: Frontiers of ferroelectricity. Springer, Berlin, pp 31–52
20. Gumennik A, Kurzweil-Segev Y, Agranat AJ (2011) Electrooptical effects in glass forming liquids of dipolar nano-clusters embedded in a paraelectric environment. Opt Mater Express 1(3):332–343
21. Hofmeister R, Yagi S, Yariv A, Agranat AJ (1993) Growth and characterization of kltn: Cu, v photorefractive crystals. J Cryst Growth 131(486):137
22. Gumennik A, Agranat AJ, Shachar I, Hass M (2005) Thermal stability of a slab waveguide implemented by α particles implantation in potassium lithium tantalate niobate. Appl Phys Lett 87(25):251917
23. Gumennik A, Ilan H, Fathei R, Israel A, Agranat AJ, Shachar I, Hass M (2007) Design methodology of refractive index engineering by implantation of high-energy particles in electro-optic materials. Appl Opt 46(19):4132–4137
24. Chang Y-C, Wang C, Yin S, Hoffman RC, Mott AG (2013) Giant electro-optic effect in nanodisordered KTN crystals. Opt Lett 38(22):4574–4577
25. Chang Y-C, Wang C, Yin S, Hoffman RC, Mott AG (2013) Kovacs effect enhanced broadband large field of view electro-optic modulators in nanodisordered KTN crystals. Opt. Express 21(15):17760–17768
26. Parravicini J, Conti C, Agranat AJ, DelRe E (2012) Programming scale-free optics in disordered ferroelectrics. Opt Lett 37(12):2355–2357
27. Barak A, Peleg O, Stucchio C, Soffer A, Segev Mordechai (2008) Observation of soliton tunneling phenomena and soliton ejection. Phys Rev Lett 100(15):153901
28. Linzon Y, Morandotti R, Volatier M, Aimez V, Ares R, Bar-Ad S (2007) Nonlinear scattering and trapping by local photonic potentials. Phys Rev Lett 99(13):133901
29. Peccianti Marco, Dyadyusha Andriy, Kaczmarek Malgosia, Assanto Gaetano (2008) Escaping solitons from a trapping potential. Phys Rev Lett 101(15):153902
30. Chen Z, Segev M, Christodoulides DN (2012) Optical spatial solitons: historical overview and recent advances. Rep Prog Phys 75(8):086401
31. Efremidis NK, Hizanidis K (2008) Disordered lattice solitons. Phys Rev Lett 101(14):143903
32. Pierangeli D, Di Mei F, Conti C, Agranat AJ, DelRe E (2015) Spatial rogue waves in photorefractive ferroelectrics. Phys Rev Lett 115(9):093901
33. Pierangeli D, Flammini M, Di Mei F, Parravicini J, de Oliveira CEM, Agranat AJ, DelRe E (2015) Continuous solitons in a lattice nonlinearity. Phys Rev Lett 114(20):203901
34. Trillo S, Torruellas W (2001) Spatial Solitons. Springer, Physics and astronomy online library. ISBN 9783540416531, https://books.google.it/books?id=_fmHJVruaogC
35. Eisenberg HS, Silberberg Y, Morandotti R, Aitchison JS (2000) Diffraction management. Phys Rev Lett 85(9):1863
36. Firstenberg Ofer, London Paz, Shuker Moshe, Ron Amiram, Davidson Nir (2009) Elimination, reversal and directional bias of optical diffraction. Nat Phys 5(9):665–668
37. Kosaka Hideo, Kawashima Takayuki, Tomita Akihisa, Notomi Masaya, Tamamura Toshiaki, Sato Takashi, Kawakami Shojiro (1999) Self-collimating phenomena in photonic crystals. Appl Phys Lett 74(9):1212–1214
38. Staliunas Kestutis, Herrero Ramon (2006) Nondiffractive propagation of light in photonic crystals. Phys Rev E 73(1):016601

Chapter 9
Rogue Waves: Transition to Turbulence and Control Through Spatial Incoherence

Rogue waves are anomalously large amplitude phenomena developing suddenly out of normal waves, living for a short time and appearing with a probability much larger than expected from ordinary wave-amplitude statistics. These extreme events have been originally observed in ocean surfaces [1] and, later on, were observed in other physical contexts, like acoustic [2] and optical dynamics [3]. They also attract interest because of their "long-tail" statistics that allow to observe events with giant amplitudes that would otherwise be truly rare and unobservable in systems that follow standard distributions. Several physical ingredients underlying the occurrence of long-tail statistics have been identified, such as interacting coherent structures emerging form stochastic instabilities [4–9], interference for quasi-random wave fields [10–12], wave-turbulence in incoherent nonlinear propagation [2, 13–17] and spatiotemporal chaos in dissipative and cavity dynamics [18–23]. Extreme amplitude events in hydrodynamic, acoustic and optical wave dynamics, have been shown to present common features also when different physical mechanisms are involved in their generation [24]. In nonlinear beam propagation, abnormal waves have been recently shown to emerge [25] due to the combination of disorder and giant electro-optic response typical of disordered photorefractive crystals in proximity of ferroelectric transition. However, fundamental issues such as the role of wave disorder and input field incoherence, remain open.

The contents of this chapter are published in *'Turbulent Transitions in Optical Wave Propagation'* Physical Review Letters **117**, 183902 (2016) and *'Enhancing optical extreme events through input wave disorder'* Physical Review A **94**, 063833 (2016).

In the Sect. 9.1 we report the direct observation of the onset of turbulence in propagating one-dimensional optical waves. The transition occurs as the disordered hosting material passes from being linear to one with extreme nonlinearity. As response grows, increased wave interaction causes a modulational unstable quasi-homogeneous flow to be superseded by a chaotic and spatially incoherent one. Statistical analysis of high-resolution wave behavior in the turbulent regime

G. Di Domenico, *Electro-optic Photonic Circuits*, Springer Theses,
https://doi.org/10.1007/978-3-030-23189-7_9

unveils the emergence of concomitant rogue waves. The transition, observed in a photorefractive ferroelectric crystal, introduces a new and rich experimental setting for the study of optical wave turbulence and information transport in conditions dominated by large fluctuations and extreme nonlinearity. In the Sect. 9.2 we investigate the role of the spatial coherence scale on the emergence of extreme events in spatially-extended photorefractive propagation tailoring the spatial autocorrelation of a partially-incoherent input field. In particular, we report a scale-dependent behavior of the long-tail statistics, which are greatly enhanced for a specific scale of the spatial incoherence. Remarkably, high-resolution measurements of the rogue waveforms reveal that these form with a characteristic intensity-independent size that coincides with the enhancing spatial incoherence scale. Using a photorefractive soliton-based model, this suggests a principal role played by saturation.

9.1 Turbulent Transitions in Optical Wave Propagation

Turbulence is a universal phenomenon in which a system is characterized by many out-of-equilibrium degrees of freedom [26]. Turbulent transitions attract great interest because the onset of spatio-temporal disorder profoundly changes the physical features of a system, the paradigm being the transport and drag properties of a fluid in a pipe and channel flow [27–29]. Manifestations of turbulence can also occur in waves, these including acoustic [2], spin [30] and optical waves [31]. In fact, when nonlinear interaction involves the excitation of a large number of waves, phase and amplitude fluctuations may lead to a stochastic field described statistically using wave turbulence theory [32]. Wave turbulence usually refers to weakly nonlinear wave systems in which the linear evolution scale can be separated from the nonlinear one. Generally these systems are dominantly influenced by some external noise and have negligible intrinsic (internal) disorder. On the other hand, as linear and nonlinear scales are comparable, strongly nonlinear coherent structures may emerge and interplay with the incoherent wave field (strong wave turbulence). In optics the onset of strong turbulence greatly alters coherence and statistics of light, as observed for pulse trains in a ring resonator [33], semiconductor lasers with feedback [34], and, recently, in tailored Raman fiber lasers [35–38]. However, experimental studies of wave turbulent behavior in the spatial domain, where light is not trapped and actually propagates in space, are especially challenging [39–42]. In particular, direct evidence of a fully-developed turbulent transition for propagating waves has remained elusive.

Here we observe the onset of turbulence for nonlinear beam propagation in photorefractive ferroelectric crystals. In our experiments, for strong nonlinearity, one-dimensional wave dynamics sharply pass from a coherent and homogeneous state to a fully-incoherent and disordered one. The transition involves the coupling of two stochastic effects: external noise associated to the initial condition and internal fluctuations of the nonlinear response. For coherent and quasi-homogeneous initial states optical turbulence set in via modulational instability, with the appearance of stochastic features and shot-to-shot fluctuations. In the turbulent regime, where a large number of spatial spectral modes are found to compete and interplay, we also identified the emergence of transient optical rogue events may be supported by different mechanisms [4, 8, 9, 18, 21, 43], optical turbulence among these [13–15, 17]. In this respect, our results provide clear evidence of the specific role of both noise-seeded instability and wave-turbulent dynamics in generating rogue waves for spatially-extended nonlinear beam propagation.

9.1.1 Characterization of Transverse Breaking

To reveal transitions to turbulence in the spatial domain we make use of an experimental setup Fig. 9.1a based on the unique out-of equilibrium photorefractive and electro-optic properties of nanodisordered ferroelectric crystals in proximity of the structural phase transition [44, 45], which has been shown to support a very rich nonlinear light dynamics [46–48]. A line (one-dimensional) gaussian beam (wavelength $\lambda = 532\,\text{nm}$) of waist $\omega_0 = 7\,\mu\text{m}$ along the x-direction and quasi-homogeneous along the y-direction is launched in a photorefractive ferroelectric crystal of potassium-lithium-tantalate-niobate (KLTN), $K_{1-\alpha}Li_\alpha Ta_{1-\beta}Nb_\beta O_3$, with $\alpha = 0.04$ and $\beta = 0.38$. The sample is a zero-cut optical quality specimen with size $2.4^{(x)} \times 2.0^{(y)} \times 1.7^{(z)}$ mm ($l_x \times l_y \times l_z$) and with the structural transition occurring at the Curie temperature $T_C = 294\,\text{K}$; large dielectric fluctuations generally persist also above this point so that nonlinear light dynamics is studied systematically with high accuracy at $T = T_C + 2K$. The input wave copropagates along the z-axis of the crystal with an uniform background intensity and nonlinearity sets in when an external bias field is applied parallel to the polarization of the propagating wave (maximum electro-optic coupling). The spatial intensity distribution is measured at the input and then at the output of the crystal along the initially quasi-homogeneous y-direction in different nonlinear conditions by means of an high-resolution imaging system composed by an objective lens ($NA = 0.5$) and a CCD camera at 15 Hz.

As a physical parameter to study the transition to turbulence we consider the physical time ruling light dynamics at the crystal output. In fact, the photorefractive nonlinearity has the peculiar property of being noninstantaneous and accumulates in time, since it involves a build-up of a photogenerated space-charge field [49]. In this way, observations at different times correspond to beam propagation for increasing nonlinearity up to saturation [50]. A typical time scale τ for beam dynamics is fixed through its symmetry-breaking into periodic coherent structures Fig. 9.1a, a process that inhibits stable spatial (1+1)D soliton formation. In fact, this stage can be accurately identified experimentally and we first characterize it varying the accessible experimental parameters. In particular, changing the input power, we measure the threshold voltage to observe transverse break-up and filaments formation. Results are reported in Fig. 9.1b and show how, increasing the input power, an almost linear scaling is found. Moreover, fixing the bias field to $V = 500\,\text{V}$ and varying the input power, we measure the averaged time τ, that is the effective nonlinearity at which the periodic break-up is observed. τ is found to decrease also linearly with the input power

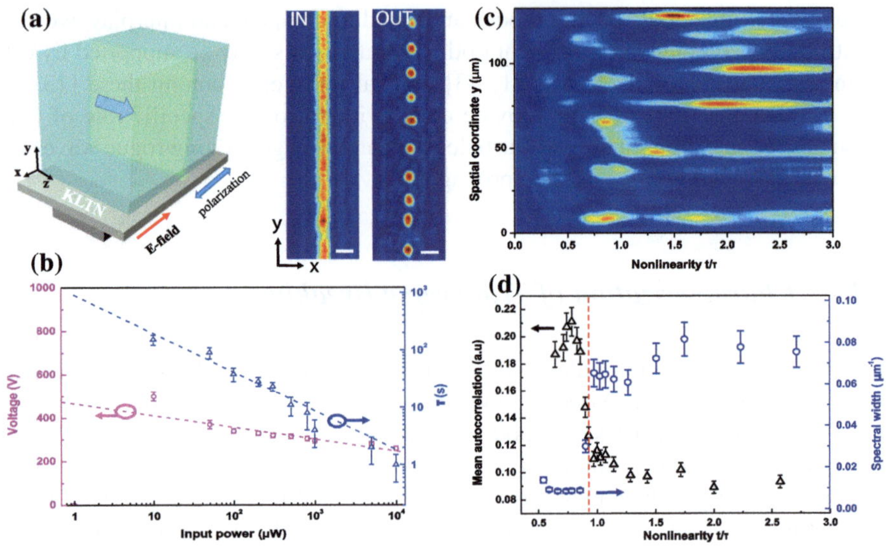

Fig. 9.1 Nonlinear wave propagation in unstable photorefractive ferroelectric crystals. **a** Sketch of the setup geometry adopted. Scale bars for input and output intensity distributions correspond to 20 μm. **b** Characterization of transverse breaking: intensity dependence of the process, with the minimum required voltage and the average time scale τ providing the formation of periodic structures. Lines are linear fits. Observation of the turbulent transition in one-dimensional beam dynamics. **c** Detected output spatial intensity distributions as a function of the nonlinearity expressed through the continuous dimensionless control parameter t/τ. **d** Corresponding width of the spatial Fourier spectrum and mean intensity autocorrelation increasing the nonlinearity. Red line serves as guide at the sharp transition signaling the onset of optical turbulence

Fig. 9.1b, in agreement with the fact that the photorefractive nonlinearity build up rate is inversely related to the peak intensity [49]. The dimensionless continuous control parameter of the nonlinearity is thus t/τ, where t is the evolving time. Hereafter we consider a laser power $P = 0.5\,\mu$W, with $\tau \approx 8$ s Fig. 9.1b. We estimate local variations of the refractive index up to 10^{-3} at $t/\tau \simeq 1$ and up to 10^{-2} for $t/\tau \simeq 2$.

9.1.2 Evidence of Turbulent Transitions

Direct evidence of the onset of turbulence as the nonlinearity increases is reported in Fig. 9.1c, d. Once that the quasi-homogeneous input line beam has experienced symmetry-breaking via modulational instability, a sharp transition into a chaotic state with pseudo-recurrent patterns occurs for $t/\tau \gtrsim 1$ Fig. 9.1c. Some of these filaments can have an extremely large intensity, as we discuss hereafter. This transition corresponds to the loss of spatial coherence that persists only on small scales. Measuring the width of the spatial Fourier spectrum, we found a sharp increase of almost one order of magnitude Fig. 9.1d. Correspondingly, the long-range autocorrelation of the intensity light distribution $I(y)$ abruptly decreases, as shown in Fig. 9.1d, where we have averaged over large r distances the absolute value of the quantity (autocorrelation function) $g(r) = \langle [I(y) - \langle I \rangle] \times [(I(y+r) - \langle I \rangle] \rangle$ normalized to $g(0)$. We stress that detecting the onset of turbulence as a sharp transition signalling departure from coherence is a method that goes beyond the stability properties of the input flux. In fact, an analogous transition has been reported in conditions where the homogeneous state of the dynamics is stable [37] and unstable [38] with respect to perturbations.

We note that a behavior similar to Fig. 9.1 has been numerically observed studying the nonlinear stage of modulation intability in the framework of the nonlinear Schroedinger Equation [51]. Here, the incoherent state generated during wave evolution is referred as integrable turbulence [51]. However, our results depart from this scenario since the presence of a saturable nonlinearity makes wave dynamics non-integrable. This means that the observed turbulent transition weakly depends on the input wave and can occur also without the modulational instability process. We demonstrate this repeating the experiments with an inhomogeneous coherent input wave; using a spatial light modulator (SLM) the input field is modulated along the y-direction with a periodic component. As shown in Fig. 9.2a a transition to turbulence is observed at $t/\tau \approx 1$. In this case, beam breaking is dominated by the input spatial frequency $k_y = 0.02\,\mu\text{m}^{-1}$ and modulation instability is only weakly involved, as noise experiences small amplification on this scale. In Fig. 9.2b the input power spectrum is shown in comparison with the quasi-homogeneous case. The picture can be easily extended to generically modulated input waves.

9.1.3 Statistical Properties of Optical Events

In order to study statistical and stochastic properties of the optical state before and after the transition to turbulence, we consider the quasi-homogeneous input case and we collect data for approximately two hundred uncorrelated experiments in the same conditions used in Fig. 9.1. Each realization naturally presents a different noise configuration, which is caused by fluctuations of the input wave arising from the experimental setup Fig. 9.2b and by local variations of the electro-optic response. These two stochastic effects are coupled, since local intensity fluctuations are amplified by the giant response of the material and inhomogeneity in the nonlinearity strongly affects light dynamics. We underline that fast material fluctuations are crucial in observing the onset of turbulence; the transition is found to disappear as the crystal is heated to a few degrees above the operational temperature. Moreover, since disorder in the material is not fixed on the time-scale of the experiment and it is furthermore modified by the wave, Anderson localization effects cannot occur in our case [52, 53]. An ensemble spectral analysis at moderate nonlinearity preceding the transition reveals the modulational unstable regime. The well-defined peak in Fig. 9.2f shows that the typical spatial frequency experiencing maximum gain is $\bar{k}_y = 0.05\,\mu\text{m}^{-1}$. However, during the single-shot dynamics, higher/lower

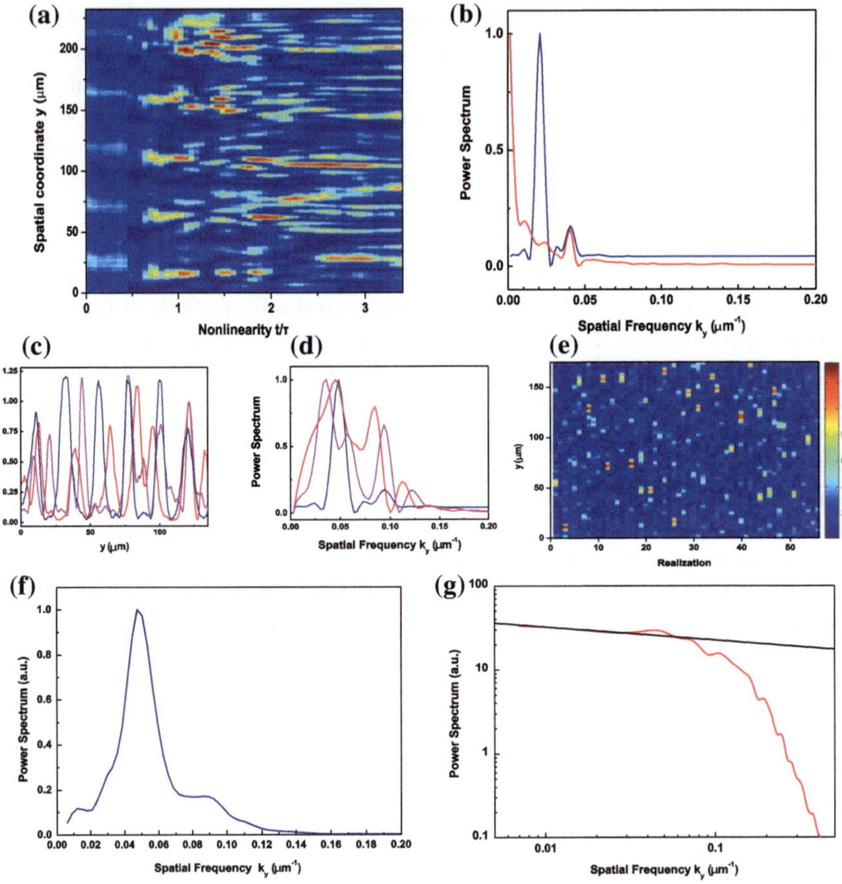

Fig. 9.2 Onset of optical turbulence for spatially-modulated input waves and spectral properties of the optical state before and after the turbulent transition. **a** Output intensity distribution increasing the nonlinearity t/τ. **b** Input power spectrum for spatially-modulated beams (blue line) in comparison with the quasi-homogeneous case of Fig. 9.1 (red line). **c–d** Intensity and spectral sample distributions of single-shot measurements at moderate nonlinearity ($t/\tau \approx 1$) showing the excitation of the spatial frequency $\bar{k}_y = 0.05\,\mu\text{m}^{-1}$ (blue line), spectrally-broad noise amplification (red line) and simultaneous development of well-defined low and high frequency modes (magenta line). **e** Single-shot disordered intensity distributions detected in the turbulent regime at $t/\tau \simeq 2.5$. **f** Ensemble spectrum of modulational instability before the transition ($t/\tau \approx 1$). **g** Measured wave-turbulent power spectrum (red line) fitted on large spatial scales with the scaling behavior $\propto \bar{k}_y^{-\gamma}$, $\gamma = 0.15 \pm 0.01$ (black line)

frequencies can also emerge easily and compete with the characteristic one. In fact, modulational instabilities are generally known to posses a strong dependence on the specific noise-realization, with properties varying from shot-to-shot [54]. In Fig. 9.2c, d we show single-shot measurements, each as an example characterizing a particular type of fluctuation. We note that as the frequency $\bar{k}_y$ is mainly excited, localized structures have a weakly-varying peak intensity and there is equipartition of power across the generated mode (see also inset in Fig. 9.1a). On the other hand, broad and double-frequency amplification results into a coherent pattern presenting large intensity fluctuations. Completely different is the scenario in the turbulent regime. Intensity distributions vary stochastically from shot-to-shot, as shown in Fig. 9.2e for several independent realizations acquired at $t/\tau \simeq 2.5$, where we expect the non-linearity to be fully saturated. Waves are characterized by random phases in analogy with optical realizations of wave turbulence theory [39], although from the statistics discussed hereafter we realize that some correlations between modes actually exist. In Fig. 9.2g we report the ensemble power spectrum; it is extremely broad and without specific resonances, with the peak associated to the amplification of $\bar{k_y}$ before the transition that results fully relaxed towards lower spatial frequencies. The spectrum results well-fitted at low frequencies by a power law behavior $\propto k_y^{-\gamma}$, with the scaling exponent $\gamma = 0.15 \pm 0.01$. Therefore, we observe evidence of an inverse cascade as the nonlinearity increases, since the majority of the wave action is now located at low transverse wavenumbers. However, this flux of wave action towards large scales should be distinguished from the one occurring in wave turbulence theory. In weak turbulence, an inverse cascade occurs for random waves at weak nonlinearity under forcing at intermediate scales [39, 40]; here, it occurs in highly nonlinear conditions and after the modulational instability stage.

In the disordered regime, part of which is shown in Fig. 9.2e, we also note the appearence of several bright localized spots. Statistical analysis shows that they are rogue waves. We first consider the probability distribution function (PDF) of peak-intensity values of localized structures emerging from instabilities before the turbulent transition. We analyze more than 10^3 events, so as to populate the histogram reported in Fig. 9.3a. This PDF contains a high-intensity peak embedded into a broad distribution. The peak, at $I/<I>\simeq 1.2$, is deterministic and closely related to the peak in the gain spectrum of Fig. 9.2f, as it arises from structures belonging to the maximally amplified frequency $\bar{k_y}$ (see Fig. 9.2c). Random fluctuations in this

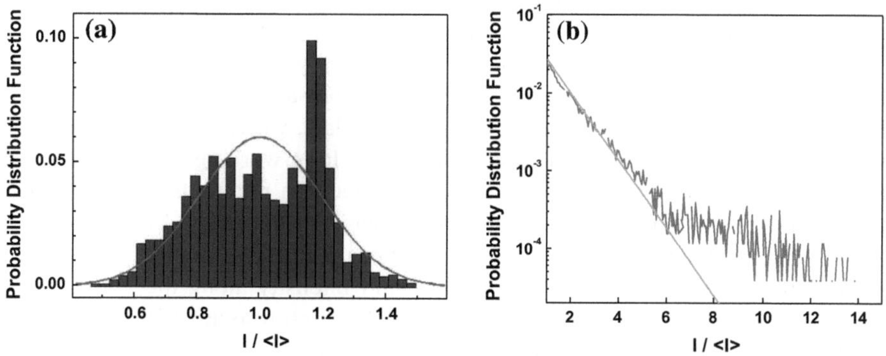

Fig. 9.3 Spatial rogue wave generation in the turbulent regime. **a** Peak-intensity PDF of localized structures emerging from instability at $t/\tau \approx 1$, experimental counts (blue bars) and gaussian trend of the distribution tails (red line). **b** Measured long-tail statistics in optical turbulence at $t/\tau \approx 2.5$ (blue line) and consistent gaussian exponential scaling for comparison

stage populate the rest of the distribution, with tails compatible with a gaussian decay, implying that extreme events occur here with low probability. This allows us to conclude that in our system giant perturbations not arise in coherent structures generated by stochastic fluctuations in instability. On the contrary, as reported in Fig. 9.3b, the PDF measured deep into the turbulent regime at $t/\tau \simeq 2.5$ presents the long-tail anomalous behavior defining rogue wave phenomena. In fact, for large intensities it deviates from the gaussian distribution expected for incoherent fields, that implies a decay according to $P(I) = \exp(-I/\langle I \rangle)/\langle I \rangle$ [55]. We note that our setup is able to detect with high-resolution the formation dynamics of each rogue wave. We found that extreme events suddenly disappear as t/τ further varies, so that no traces are found after their passing. Moreover, from our data, inelastic interactions between less intense structures in the wake of extreme events are not so evident. This fact may involve the presence of a different saturation-dependent process in rogue wave appearence.

9.2 Enhancing Optical Extreme Events Through Input Wave Disorder

Here we investigate how the emergence of extreme events strongly depends on the correlation length of the input field distribution. We observe the behavior of optical waves in turbulent photorefractive propagation with partially-incoherent excitations, we find that rogue waves are strongly enhanced for a characteristic input correlation scale. Waveform analysis identifies this scale with a characteristic peak-intensity-independent wave size, suggesting a general role played by saturation in the nonlinear response in rogue phenomena.

In our experiments we make use of partially-incoherent beams propagating in photorefractive ferroelectrics, where rogue events have been observed through coherent one-dimensional input excitations [25]. Setup and methods are shown schematically in Fig. 9.4a. They are based on the peculiar electro-optic features of disordered ferroelectric crystals in proximity of their structural phase transition [45, 47, 48, 56, 57] and on the photorefractive propagation of partially-incoherent beams [58, 59]. Specifically, light at a wavelength $\lambda = 532$ nm from a 150 mW continuous-wave laser is expanded and focused on a glass diffuser plate, where transmitted radiation is collected producing a collimated speckle field. A long-working-distance objective ($NA = 0.55$) launches this field at the input facet of a photorefractive ferroelectric crystal of potassium-lithium-tantalate-niobate (KLTN), $K_{1-\alpha}Li_{\alpha}Ta_{1-\beta}Nb_{\beta}O_3$ ($\alpha = 0.04, \beta = 0.38$). The partially-incoherent beam, linearly polarized in the experimental plane, copropagates with a background intensity along the z-axis of the crystal and is detected at the output facet through a high-resolution imaging system ($NA = 0.50$) and a CCD camera. The sample is a zero-cut optical quality specimen with size $2.4^{(x)} \times 2.0^{(y)} \times 1.7^{(z)}$ mm ($l_x \times l_y \times l_z$) and with the ferroelectric transition occurring at the Curie temperature $T_C = 294$ K. Spatiotemporal fluctuations of the media response persist also slightly above this point so that disorder-affected light

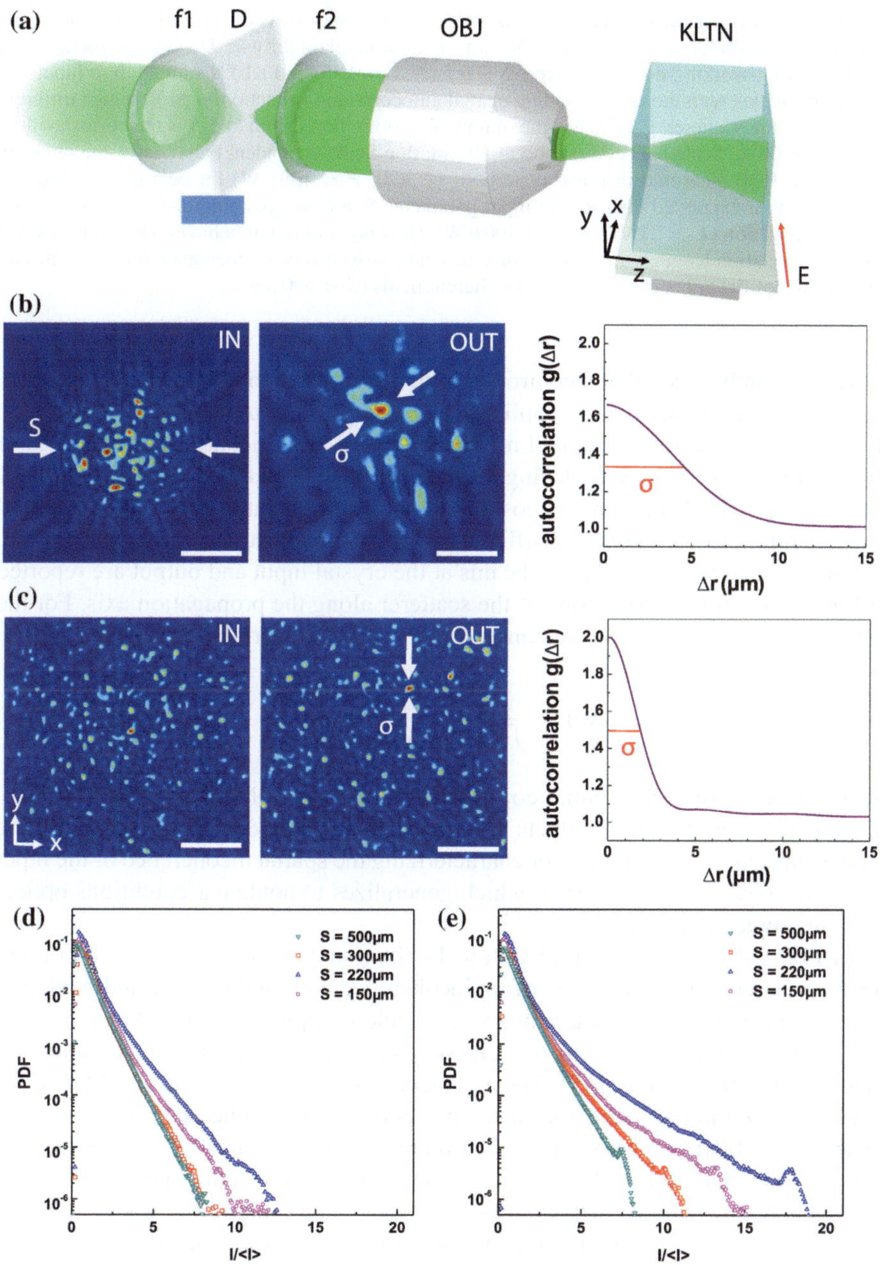
(a)
f1
D
f2
OBJ
KLTN
y
x
z
E
(b)
IN
S
OUT
σ
autocorrelation g(Δr)
Δr (μm)
(c)
IN
OUT
σ
autocorrelation g(Δr)
Δr (μm)
(d)
S = 500μm
S = 300μm
S = 220μm
S = 150μm
PDF
I/<I>
(e)
S = 500μm
S = 300μm
S = 220μm
S = 150μm
PDF
I/<I>

◂**Fig. 9.4** Partially-incoherent beams in photorefractive ferroelectric crystals. **a** Sketch of the experimental setup with lenses ($f1 = f2 = 50$ mm), adjustable glass diffuser D (average particle size of 2 μm), long-working-distance objective OBJ ($NA = 0.55$) and KLTN sample. **b–c** Input and output intensities with the corresponding spatial autocorrelation function $g(\Delta r)$ for two different positions of the scatterer. σ indicates the output autocorrelation length and S is the corresponding input source size. Scale bars correspond to 30 μm. **d–e** Scale-dependent behavior of the intensity statistics. **d** PDF measured in linear conditions ($E = 0$, $P = 400\,\mu$W) for beams with different coherent length expressed through the input parameter S. **e** Corresponding distributions for nonlinear propagation ($E = 2\,$kV/cm, $P = 400\,\mu$W), showing a long-tail behavior depending on the specific correlation length, with a large enhancement in rogue waves appearance for $S = 220\,\mu$m. Suppression of the tail occur for highly-incoherent fields ($S = 500\,\mu$m)

dynamics can be studied and controlled with high reproducibility at $T = T_C + 2K$. Here, an external bias field is applied transversely to the propagation axis to tune the strength of the Kerr-saturated nonlinearity [49]. The incoherence properties of the input beam are achieved placing the diffuser inbetween two confocal lenses ($f1$ and $f2$) and varied changing its position along the propagation axis, whereas small tilts and rotations on it generate different disordered realizations of the optical field. Examples of partially-incoherent beams at the crystal input and output are reported in Fig. 9.4b, c for two positions of the scatterer along the propagation axis. For the output intensity distribution we consider the spatial autocorrelation function

$$g(\Delta r) = \frac{\langle \int d^2r I(r) I(r + \Delta r)\rangle}{\int d^2r \langle I(r)\rangle\langle I(r + \Delta r)\rangle}, \tag{9.1}$$

whose width defines the spatial correlation length σ, i.e. the average speckle size. Since σ varies as nonlinear effects are involved in wave dynamics [60], we use the input source size S as a parameter characterizing the spatial incoherence of the input beam. We have $S \simeq 2\lambda l_z/g(0)\sigma$, which generalizes to nonlinear conditions optical speckle propagation [60, 61].

In Fig. 9.4 the detected probability distribution function (PDF) of the output intensity is reported varying the beam incoherence both in the linear and nonlinear case. In linear conditions, where no external field is applied, in Fig. 9.4d we observe no significant deviations from the gaussian statistics as expected for completely-random interfering waves [55]. The exponential scaling $PDF = \exp(-I/\langle I\rangle)/\langle I\rangle$ is well verified in particular for beams presenting spatial coherence only on small scales ($S \approx 500\,\mu$m, $S \approx 300\,\mu$m). For more correlated beams ($S \approx 220\,\mu$m, $S \approx 150\,\mu$m), the PDF slightly deviates at large intensities, consistently with the presence of weak inhomogeneities in the phases of the elementary interfering waves [10, 12]. Rogue waves occur as the nonlinearity is activated by means of the external field $E = 2\,$kV/cm. In the nonlinear case, the incoherent field experiences strong self-interaction and spatiotemporal fluctuations so that we observe the speckle intensity dynamically varying in a turbulent fashion [57]. To study the statistics in this stage, we acquire more than two hundred independent spatial distributions for a fixed 400 μW input power and sample conditions. Results as a function of the coherence length are shown in Fig. 9.4e and demonstrate how extreme events strongly depend from this parameter. We found the long-tail statistics defining rogue waves and a peculiar scale-dependent behavior. Specifically, the spatial correlation scale of the optical field strongly affects its PDF, with a large enhancement in extreme event appearance that occurs for incoherent beams of size $S \approx 220\,\mu\text{m}$ and their complete suppression at $S \approx 500\,\mu\text{m}$. We note that the effect is approximately independent of the input power and of the value of the applied field, provided that both are above a certain threshold ensuring the highly-nonliner dynamics. Therefore, we observe that small-scale random intensity fluctuations inhibit rogue wave generation, whereas a peculiar increase in their probability is triggered by a specific input disorder scale. We note that a similar inhibition for highly-incoherent waves has been also reported in the temporal turbulent dynamics of passive optical fiber ring cavity [13].

To investigate the mechanism underlying the correlation between abnormal wave statistics and incoherence scale, we use our ability to resolve the spatial waveform of each event with $0.3\,\mu\text{m}$ resolution (for typical wave features of $10\,\mu\text{m}$). We first consider the data set with incoherence corresponding to the maximum statistical-tail enhancement and, in particular, we analyze the rogue wave peak intensity I_P and its full-width-at-half-maximum ΔX. Examples of spatially-resolved rogue waveforms emerging from partially-incoherent intensity distributions are shown in Fig. 9.5b as giant pulses. In Fig. 9.5a we report an interesting behavior that is found for the two analyzed parameters: even though the abnormal waves span different peak intensities, their width is almost constant. Localized events appear with the same transverse size

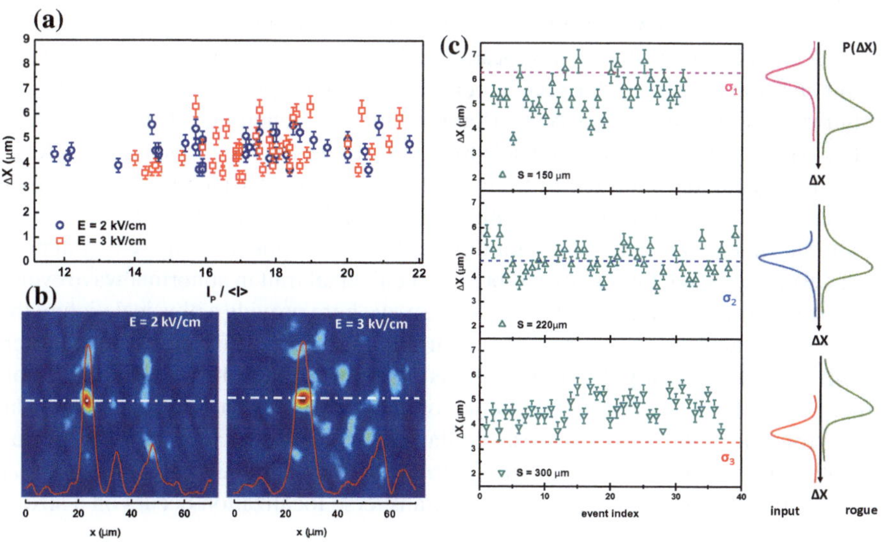

Fig. 9.5 Unveiling optical rogue waveforms. **a** Detected transverse width ΔX and peak intensity I_P of extreme events for data at different applied fields. **b** High-resolution spatial intensity distributions containing localized abnormal waveforms. Red curves are x-profiles along the dotted lines. **c** Evidence of a typical scale in rogue waveforms. Measured extreme events widths at different coherence length σ (dashed lines). The two scales are resonant for $S = 220\,\mu\text{m}$, where a large increase results in the probability of rogue wave appearance see Fig. 9.4e. The diagram on the right illustrates how results imply the presence of a typical spatial scale for rogue events

irrespective of the fact that they populate the gaussian portion of the PDF or the extreme one of its abnormal tail. This feature persist also at different bias fields and, as further shown in the following, it amounts to a general property of rogue waves in the saturable nonlinearity. Therefore, we extend our analysis taking into account the width of the extreme events as a function of the degree of incoherence of the corresponding optical field. Specifically, we compare their typical scale ΔX with the correlation scale σ of its entire intensity distribution, obtained in the nonlinear regime according to Eq. (9.1). This allows us to inspect whether the rogue wave has a size determined by the mean autocorrelation length of the speckle beam or an intrinsic properties is involved. The whole picture is presented in Fig. 9.5c. Extreme events are found to emerge on a typical scale that is significantly lower or higher than the coherence one, respectively, for beams of size $S = 150\,\mu\text{m}$ and $S = 300\,\mu\text{m}$ see Fig. 9.4e. Moreover, matching between these two scales is evident at $S = 220\,\mu\text{m}$, that is exactly the case in which the large enhancement in the long-tail statistics is detected. The findings prove that the key feature providing the optimal input disorder conditions for the emergence of non-Gaussian statistics is the existence of an intrinsic scale for rogue waves. We estimate it to approximately $\overline{\Delta X} = 4.5\,\mu\text{m}$. In fact, as schematically illustrated in Fig. 9.5c, the coherence length distribution of the input beam acts as a probe for the probability $P(\Delta X)$ of finding extreme events with a certain width ΔX. Their overlap, in terms of sizes, sets the amount of emerging extreme events, so that the enhancement at $S = 220\,\mu\text{m}$ appears as a resonant interaction.

The existence of a preferential size for extreme waves provides a key information to understand statistically their appearance. Moreover, once the nonlinear propagation conditions are fixed, the spatial correlation of the optical field can be tuned to arbitrarily modify the intensity distribution tail. The generality of this mechanism relies on the physical basis that leads to a typical size for rogue events. We address the fundamental question on its origin starting from the consideration that the main properties of the photorefractive nonlinearity underlying our optical dynamics is its saturable character. Since saturation turns out in the response of any real system for large excitations, the finding may represent a universal trait in abnormal wave events, at least in this limit condition. For our system, we here provide a physical picture that not only explains the presence of a peculiar spatial scale, but also the observed insensibility to wave intensity. The framework is based on spatial solitons in saturbale nonlinearities, whose structural and interaction properties have been suggested to play a key role for rogue waves in these media [25]. We consider their non-equilibrium counterpart, i.e., transient self-trapping waves in non-stationary conditions [62]. As detailed in Ref. [63], in the present case, transverse localization occurs on a size

$$\Delta x \simeq \frac{3\lambda}{2\pi n^2 a_{eo}} E^{-2}, \tag{9.2}$$

where n is the linear index of refraction and a_{eo} a parameter quantifying the electro-optical response of the media. For our experimental realization, we have $\Delta x = 5 \pm 1\,\mu\text{m}$, where the uncertainty is related to the uncertainty in a_{eo} in proxim-

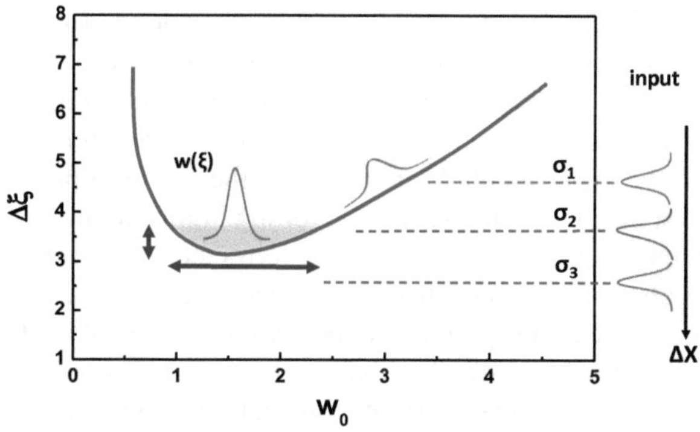

Fig. 9.6 Possible mechanism underlying the appearance of extreme intensity fluctuations. Existence curve of non-equilibrium solitons in saturable nonlinearities (red line) on which the waveform $w(\xi)$ is schematically shown. Arrows indicate the magnitude of width and amplitude fluctuations in the gray region, that is in proximity of the localized, self-trapped, wave solution. For comparison with Figs. 9.5c and 9.4, input correlation lengths used in experiments are also reported

ity of the ferroelectric phase transition for biased condition [56]. This value of Δx is consistent with the typical scale of rogue waveforms $\overline{\Delta X}$ we have found. Moreover, Eq. (9.2) possess the fundamental property of being independent on the wave intensity, in agreement with our observations of extreme events Fig. 9.5a. Dependence on the external electric field E is predicted as very weak at high values and the result in Fig. 9.5a with different bias fields verifies also this feature. Therefore, observations strongly suggest that the scale dependence of long-tail statistics with spatial incoherence can be explained with the mechanism illustrated in Fig. 9.6, where the phase-space of the nonlinear waves in terms of normalized amplitude w_0 and width $\Delta\xi$ is recalled. Non-equilibrium self-trapped waves form across the minimum of the existence curve according to Eq. (9.2) [63]. Here, a small variation in $\Delta\xi$ can lead to large fluctuations of the wave amplitude, with peak intensities reaching the giant values that populate the extreme regions of the total PDF. Extreme events are enhanced when the input coherence scale falls in this region, whereas their excitations and suppression implies, respectively, that matching with the input autocorrelation is partial or does not occur at all.

9.3 Final Remarks

Summarizing in Sect. 9.1, we have observed the onset of turbulence in light flow through a nonlinear medium. The transition in one-dimensional beam dynamics occurs through nonlinear wave interaction and leads from a modulational unstable to a chaotic state where optical coherence is lost and several spatial modes are simultaneously excited. We have found that the stochastic nature of the process manifests itself in a different intensity distribution function before and after the turbulent transition, where concomitant rogue events have also been detected. These observations demonstrate how optical beam propagation in stochastic and nonlinear conditions can form a tool to generate and investigate wave turbulent phenomena. In Sect. 9.2, we have experimentally investigated the role of input wave disorder in the formation of optical extreme events. Exploiting highly-nonlinear propagation of partially-incoherent beams in photorefractive ferroelectric crystals, we reveal how the occurring of abnormal events strongly depends on the coherence length of the optical field. Tuning the input spatial autocorrelation we are able to modify the long-tail statistics of the output intensity distribution from inhibition to large enhancement. In the specific case, we are able to attribute this scale-dependent property to the onset of saturation in the nonlinearity. Our results may open important routes to control and exploit disordered light transport in extreme nonlinear conditions and pave the way to the development of optical devices that exploit the huge sensitivity of chaotic states. At the same time, they represent an important step in the understanding of anomalous wave events in spatially-extended and non-integrable systems.

References

1. Müller P, Garrett C, Osborne A (2005) Rogue waves. Oceanography 18(3):66. http://dx.doi.org/10.5670/oceanog.2005.30
2. Ganshin AN, Efimov VB, Kolmakov GV, Mezhov-Deglin LP, McClintock PV (2008) Observation of an inverse energy cascade in developed acoustic turbulence in superfluid helium. Phys Rev Lett 101(6):065303
3. Solli DR, Ropers C, Koonath P, Jalali B (2007) Optical rogue waves. Nature 450(7172):1054–1057
4. Armaroli A, Conti C, Biancalana F (2015) Rogue solitons in optical fibers: a dynamical process in a complex energy landscape? Optica 2(5):497–504
5. Birkholz S, Nibbering ETJ, Brée C, Skupin S, Demircan A, Genty G, Steinmeyer G (2013) Spatiotemporal rogue events in optical multiple filamentation. Phys Rev Lett 111(24):243903
6. Dudley JM, Dias F, Erkintalo M, Genty G (2014) Instabilities, breathers and rogue waves in optics. Nat Photon 8(10):755–764
7. Lecaplain C, Grelu P, Soto-Crespo JM, Akhmediev N (2012) Dissipative rogue waves generated by chaotic pulse bunching in a mode-locked laser. Phys Rev Lett 108(23):233901
8. Onorato M, Waseda T, Toffoli A, Cavaleri L, Gramstad O, Janssen PA, Kinoshita T, Monbaliu J, Mori N, Osborne AR et al (2009) Statistical properties of directional ocean waves: the role of the modulational instability in the formation of extreme events. Phys Rev Lett 102(11):114502
9. Shats M, Punzmann H, Xia H (2010) Capillary rogue waves. Phys Rev Lett 104(10):104503
10. Arecchi FT, Bortolozzo U, Montina A, Residori S (2011) Granularity and inhomogeneity are the joint generators of optical rogue waves. Phys Rev Lett 106(15):153901
11. Höhmann R, Kuhl U, Stöckmann H-J, Kaplan L, Heller EJ (2010) Freak waves in the linear regime: a microwave study. Phys Rev Lett 104(9):093901
12. Liu C, Van Der Wel RE, Rotenberg N, Kuipers L, Krauss TF, Di Falco A, Fratalocchi A (2015) Triggering extreme events at the nanoscale in photonic seas. Nat Phys 11(4):358–363
13. Conforti M, Mussot A, Fatome J, Picozzi A, Pitois S, Finot C, Haelterman M, Kibler B, Michel C, Millot G (2015) Turbulent dynamics of an incoherently pumped passive optical fiber cavity: quasisolitons, dispersive waves, and extreme events. Phys Rev A 91(2):023823

14. Hammani K, Kibler B, Finot C, Picozzi A (2010) Emergence of rogue waves from optical turbulence. Phys Lett A 374(34):3585–3589
15. Randoux S, Walczak P, Onorato M, Suret P (2016) Nonlinear random optical waves: integrable turbulence, rogue waves and intermittency. Phys D 333:323–335
16. Suret P, El Koussaifi R, Tikan A, Evain C, Randoux S, Szwaj C, Bielawski S (2016) Single-shot observation of optical rogue waves in integrable turbulence using time microscopy. Nat Commun 7
17. Walczak P, Randoux S, Suret P (2015) Optical rogue waves in integrable turbulence. Phys Rev Lett 114(14):143903
18. Bonatto C, Feyereisen M, Barland S, Giudici M, Masoller C, Rios Leite JR, Tredicce JR (2011) Deterministic optical rogue waves. Phys Rev Lett 107(5):053901
19. Gibson CJ, Yao AM, Oppo GL (2016) Optical rogue waves in vortex turbulence. Phys Rev Lett 116(4):043903
20. Marsal N, Caullet V, Wolfersberger D, Sciamanna M (2014) Spatial rogue waves in a photorefractive pattern-forming system. Opt Lett 39(12):3690–3693
21. Montina A, Bortolozzo U, Residori S, Arecchi FT (2009) Non-gaussian statistics and extreme waves in a nonlinear optical cavity. Phys Rev Lett 103(17):173901
22. Pisarchik AN, Jaimes-Reátegui R, Sevilla-Escoboza R, Huerta-Cuellar G, Taki M (2011) Rogue waves in a multistable system. Phys Rev Lett 107(27):274101
23. Selmi F, Coulibaly S, Loghmari Z, Sagnes I, Beaudoin G, Clerc MG, Barbay S (2016) Spatiotemporal chaos induces extreme events in an extended microcavity laser. Phys Rev Lett 116(1):013901
24. Onorato M, Residori S, Bortolozzo U, Montina A, Arecchi FT (2013) Rogue waves and their generating mechanisms in different physical contexts. Phys Rep 528(2):47–89
25. Pierangeli D, Di Mei F, Conti C, Agranat AJ, DelRe E (2015) Spatial rogue waves in photorefractive ferroelectrics. Phys Rev Lett 115(9):093901
26. Landau LD, Lifshitz EM (2013) Fluid mechanics: Landau and Lifshitz: course of theoretical physics, vol 6. Elsevier, Amsterdam
27. Avila K, Moxey D, de Lozar A, Avila M, Barkley D, Hof B (2011) The onset of turbulence in pipe flow. Science 333(6039):192–196
28. Grossmann S (2000) The onset of shear flow turbulence. Rev Mod Phys 72(2):603
29. Masaki S, Keiichi T (2016) A universal transition to turbulence in channel flow. Nat Phys
30. Boyer F, Falcon E (2008) Wave turbulence on the surface of a ferrofluid in a magnetic field. Phys Rev Lett 101(24):244502
31. Picozzi A, Garnier J, Hansson T, Suret P, Randoux S, Millot G, Christodoulides DN (2014) Optical wave turbulence: Towards a unified nonequilibrium thermodynamic formulation of statistical nonlinear optics. Phys Rep 542(1):1–132
32. Nazarenko S (2011) Wave turbulence, vol 825. Springer Science & Business Media, Berlin
33. Mitschke F, Steinmeyer G, Schwache A (1996) Generation of one-dimensional optical turbulence. Phys D 96(1):251–258
34. Mork J, Tromborg B, Mark J (1992) Chaos in semiconductor lasers with optical feedback: theory and experiment. IEEE J Quantum Electron 28(1):93–108
35. Aragoneses A, Carpi L, Tarasov N, Churkin DV, Torrent MC, Masoller C, Turitsyn SK (2016) Unveiling temporal correlations characteristic of a phase transition in the output intensity of a fiber laser. Phys Rev Lett 116(3):033902
36. Turitsyn SK, Babin SA, Turitsyna EG, Falkovich GE, Podivilov EV, Churkin DV (2013) Optical wave turbulence. Adv Wave Turbul 83:113–164
37. Turitsyna EG, Smirnov SV, Sugavanam S, Tarasov N, Shu X, Babin SA, Podivilov EV, Churkin DV, Falkovich G, Turitsyn SK (2013) The laminar-turbulent transition in a fibre laser. Nat Photon 7(10):783–786
38. Wabnitz S (2014) Optical turbulence in fiber lasers. Opt Lett 39(6):1362–1365
39. Bortolozzo U, Laurie J, Nazarenko S, Residori S (2009) Optical wave turbulence and the condensation of light. JOSA B 26(12):2280–2284

40. Laurie J, Bortolozzo U, Nazarenko S, Residori S (2012) One-dimensional optical wave turbulence: experiment and theory. Phys Rep 514(4):121–175
41. Shih M-F, Jeng C-C, Sheu F-W, Lin C-Y (2002) Spatiotemporal optical modulation instability of coherent light in noninstantaneous nonlinear media. Phys Rev Lett 88(13):133902
42. Sun C, Jia S, Barsi C, Rica S, Picozzi A, Fleischer JW (2012) Observation of the kinetic condensation of classical waves. Nat Phys 8(6):470–474
43. Onorato M, Osborne AR, Serio M (2006) Modulational instability in crossing sea states: A possible mechanism for the formation of freak waves. Phys Rev Lett 96(1):014503
44. Pierangeli D, Parravicini J, Di Mei F, Parravicini GB, Agranat AJ, DelRe E (2014) Photorefractive light needles in glassy nanodisordered kntn. Opt Lett 39(6):1657–1660
45. Pierangeli D, Ferraro M, Di Mei F, Di Domenico G, De Oliveira CEM, Agranat AJ, DelRe E (2016) Super-crystals in composite ferroelectrics. Nat Commun 7:10674
46. DelRe E, Spinozzi E, Agranat AJ, Conti C (2011) Scale-free optics and diffractionless waves in nanodisordered ferroelectrics. Nat Photon 5(1):39–42
47. DelRe E, Di Mei F, Parravicini J, Parravicini G, Agranat AJ, Conti C (2015) Subwavelength anti-diffracting beams propagating over more than 1,000 rayleigh lengths. Nat Photon
48. Di Mei F, Caramazza P, Pierangeli D, Di Domenico G, Ilan H, Agranat AJ, Di Porto P, DelRe E (2016) Intrinsic negative mass from nonlinearity. Phys Rev Lett 116(15):153902
49. DelRe E, Crosignani B, Di Porto P (2009) Photorefractive solitons and their underlying nonlocal physics. Prog Optics 53:153–200
50. Qieni L, Han J, Dai H, Ge B, Zhao S (2015) Visualization of spatial-temporal evolution of light-induced refractive index in mn: Fe: Ktn co-doped crystal based on digital holographic interferometry. IEEE Photon J 7(4):1–11
51. Agafontsev DS, Zakharov VE (2015) Integrable turbulence and formation of rogue waves. Nonlinearity 28(8):2791
52. Leonetti M, Karbasi S, Mafi A, Conti C (2014) Light focusing in the anderson regime. arXiv:1407.8062
53. Segev M, Silberberg Y, Christodoulides DN (2013) Anderson localization of light. Nat Photon 7(3):197–204
54. Solli DR, Herink G, Jalali B, Ropers C (2012) Fluctuations and correlations in modulation instability. Nat Photon 6(7):463–468
55. Goodman JW (1975) Statistical properties of laser speckle patterns. In: Laser speckle and related phenomena. Springer, Berlin, pp 9–75
56. Pierangeli D, Di Mei F, Parravicini J, Parravicini GB, Agranat AJ, Conti C, DelRe E (2014) Observation of an intrinsic nonlinearity in the electro-optic response of freezing relaxors ferroelectrics. Opt Mater Express 4(8):1487–1493
57. Pierangeli D, Di Mei F, Di Domenico G, Agranat AJ, Conti C, DelRe E (2016a) Turbulent transitions in optical wave propagation. Phys Rev Lett 117(18):183902
58. Chen Z, Segev M, Christodoulides DN (2003) Experiments on partially coherent photorefractive solitons. J Opt A: Pure Appl Opt 5(6):S389
59. Mitchell M, Chen Z, Shih M, Segev M (1996) Self-trapping of partially spatially incoherent light. Phys Rev Lett 77(3):490
60. Bromberg Y, Lahini Y, Small E, Silberberg Y (2010) Hanbury brown and twiss interferometry with interacting photons. Nat Photon 4(10):721–726
61. Derevyanko S, Small E (2012) Nonlinear propagation of an optical speckle field. Phys Rev A 85(5):053816
62. Fressengeas N, Wolfersberger D, Maufoy J, Kugel G (1998) Build up mechanisms of (1+ 1)-dimensional photorefractive bright spatial quasi-steady-state and screening solitons. Opt Commun 145(1):393–400
63. DelRe E, Palange E (2006) Optical nonlinearity and existence conditions for quasi-steady-state photorefractive solitons. JOSA B 23(11):2323–2327

Chapter 10
Conclusion

The work presented reports various new findings and applications. We have demonstrated the first miniaturized modulator able to transform an input Gaussian beam into a non-diffracting Bessel-like beam using a photoinduced index of refraction pattern. This approach allows the modulator to be fast and to use low voltage, and permits in principle the integration into solid-state arrays. The observation that our modulator does not involve self-action effects is the natural starting point for studying the propagation of non-diffracting beams in the presence of non-linearity (Flammini et. al. to be submitted).

On the other hand, the study of Bessel beams in microscopy has led to the discovery of non diffracting field that possesses an effective localization degree (what we term "light droplet" field) solving the main difficulties of implementing Bessel beams in basic imaging schemes. Further results shows how our droplets can penetrate deep into biological tissues thanks to the self-healing property passed down from Bessel beams and how their use in a confocal microscope provides the z-localization. The study of the droplet field and its axial localization guide us to discover their increased radial localization. In our work, we show how the intensity of the outer ring structure of a Droplet field can be suppressed up to 1/2 with respect to the intensity of a comparable Bessel beam. At present we are testing these findings in a light-sheet microscope set-up (Di Domenico et. al. to be submetted) to demonstrate how the droplet illumination can be used to improve the field of view while keeping good sectioning of the sample.

These findings have been accompanied by experiments carried out on disordered ferroelectrics, we demonstrate how a giant diffusive photorefractive nonlinearity is able to support negative intrinsic effective mass, how the input wave disorder is involved in the formation of optical extreme events and how, in strong non-linear condition, wave dynamics sharply pass from a coherent and homogeneous state to a fully-incoherent one. We also report the observation of a new ordered ferroelectric phase characterized by the spontaneous formation of a macroscopic crystalline structure of polar domains.

In the end, from a theoretical point of view we have found an answer to the question of if and how it is possible to entangle two distant system by means of one

G. Di Domenico, *Electro-optic Photonic Circuits*, Springer Theses,
https://doi.org/10.1007/978-3-030-23189-7_10

particle. We found entanglement to be the product of the microscopic reversibility applied to a system whose number of particles increase. More particles means more independent experiments but does not mean extra information, since processes in which information is produced are forbidden by the principles of quantum mechanics. Total entanglement can only be achieved at the expense of having the two systems interact. The requirement of direct interaction between the two systems prevent the possibility of entangling them when they are already distant. The natural outcome of these arguments and, also what we are now attempting to do, is to show that single particle entanglement cannot be the basis for useful quantum technology, demonstrating so the central role of nonlinearity in the scheme for reliable quantum computing or quantum cryptography, similarly to what happens in total quantum teleportation.

Appendix
The Conditional Nature of Single Particle Entanglement

In this appendix, we show how the uncertainty principle limits the ability of a single particle to entangle two distant systems. Entanglement is shown to occur with less than 1/2 probability of success, and this only after post-selection. We also apply our analysis to a specific example generally considered capable of entangling two atoms using one photon.

A.1 General Scenario

In the general situation one particle is used to entangle two distant systems Fig. A.1a, the state of the particle p is appropriately delocalized onto two paths A and B, i.e.,

$$|p\rangle = (1/\sqrt{2})\left(|1_A, 0_B\rangle + |0_A, 1_B\rangle\right). \tag{A.1}$$

Although this state is formally entangled, it can readily be cast into a disentangled form, for example recombining the two particle paths and forming a balanced interferometer. It is congruently not capable of supporting itself two separate space-like measurements, as required for establishing nonlocality. Assume that these two paths host two separate systems, say A and B, two two-level systems (with ground state g and excited stated e) initially in their ground state $|g_A, g_B\rangle$, and that particle p is able to excite each of them separately. It would hence appear that after the interaction the initial disentangled state

$$|\psi_D\rangle = (1/\sqrt{2})(|1_A, 0_B\rangle + |0_A, 1_B\rangle)|g_A, g_B\rangle \tag{A.2}$$

becomes the entangled state

$$|\psi_E\rangle = (1/\sqrt{2})|0_A, 0_B\rangle(|g_A, e_B\rangle + |e_A, g_B\rangle). \tag{A.3}$$

G. Di Domenico, *Electro-optic Photonic Circuits*, Springer Theses,
https://doi.org/10.1007/978-3-030-23189-7

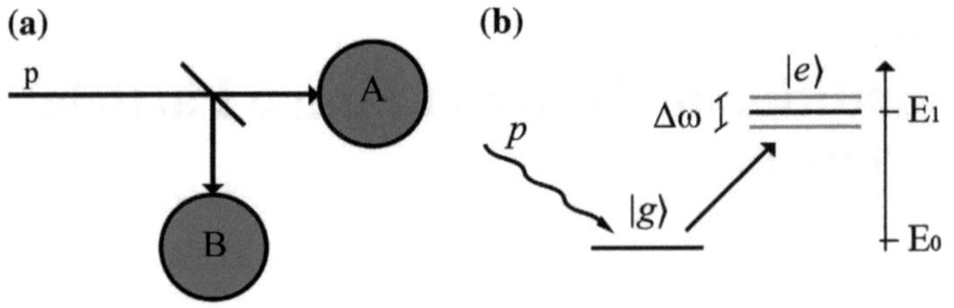

Fig. A.1 **a** General set-up: a single particle p is split onto two paths and made to interact with two distant systems A and B. **b** Energy level scheme for each particle-system interaction

This conclusion is, however, not precise. In fact, the property of A and B to be entangled is meaningful for a time interval $\Delta t \ll 1/\Delta\omega$, where $\Delta\omega$ is the linewidth of the excited level in each system Fig. A.1b. For longer time intervals, the excited states will begin to decay. On the other hand, in order for p to excite A or B, its energy uncertainty $\Delta\omega_p$ must be such that $\Delta\omega_p < \Delta\omega$. This implies that the time uncertainty in the arrival (or more generally, interaction) of particle p is $\Delta t_p > 1/\Delta\omega$. Hence, any given experiment able to test and eventually detect an entanglement between A and B requires a resolution in time less than the time uncertainty of particle p, so that the actual state is wholly uncertain between the particle p not being absorbed by system A or B or being absorbed, i.e., $|\psi\rangle = (1/\sqrt{2})(|\psi_D\rangle + |\psi_E\rangle)$, the first corresponding to the disentangled state of A and B both in the fundamental state, the second to the entangled condition. Thus, at its best, there is a maximum 1/2 probability of finding the two systems entangled.

A.1.1 Specific Example

As an example, we consider the interesting case in which p is a single photon and A and B are two two-level atoms, as described in Ref. [7]. Each atom, of transition frequency ω_a, is within a cavity able to contain photons of angular frequency ω_c such that $|\omega_a - \omega_c| \ll \omega_a + \omega_c$. The usual Jaynes–Cummings Hamiltonian [5] describing each single system atom-plus-photon reads

$$\hat{H} = \hbar\omega_c \hat{a}^+ \hat{a} + \hbar\omega_a \frac{\hat{\sigma}_z}{2} + \frac{\hbar\Omega}{2}(\hat{a}\hat{\sigma}^+ + \hat{a}^+\hat{\sigma}^-), \tag{A.4}$$

where $\hat{a}^+$ and $\hat{a}$ are the photon creation and annihilation operators and $\hbar\Omega$ denotes the amplitude of the atom-photon interaction. The characteristic angular frequency Ω can be evaluated through the relation [5]

$$\Omega^2 = \frac{4d^2\omega_a}{\hbar V \epsilon_0}, \tag{A.5}$$

where V is the cavity volume and d labels the transition dipole moment, i.e.,

$$d^2 = e^2 |\langle\psi_1|\vec{r}|\psi_0\rangle|^2, \tag{A.6}$$

$|\psi_0\rangle, |\psi_1\rangle$ being the fundamental and excited atomic states. For future use, we evaluate Eq. (A.5) for hydrogen atoms, for which

$$|\psi_0\rangle = |\psi_{n=1,l=0,m=0}\rangle = \frac{e^{-\frac{r}{a_0}}}{\sqrt{\pi a_0^3}}, \tag{A.7}$$

$$|\psi_1\rangle = |\psi_{n=2,l=1,m=0}\rangle = \frac{\frac{r}{a_0} e^{-\frac{r}{2a_0}}}{4\sqrt{2\pi a_0^3}}, \tag{A.8}$$

where $a_0 \simeq 0.52$ Å is Bohr radius. The angular frequency $\omega_a = \frac{E_1 - E_0}{\hbar}$ (E_0, E_1 are the energies of $|\psi_0\rangle, |\psi_1\rangle$) is determined through the relation $E_1 - E_0 \simeq 10.2eV$ and Eq. (A.6) is easily evaluated with the help of Eqs. (A.7), (A.8), thus obtaining

$$d^2 = \frac{2^{15}}{3^{10}} (ea_0)^2. \tag{A.9}$$

The above results and Eq. (A.5) give

$$\Omega(s^{-1}) \simeq \frac{50}{V^{\frac{1}{2}}} (V \rightarrow m^3). \tag{A.10}$$

Returning to the general case, the atom is conveniently described through the spin 1/2 formalism where the standard ladder operators $\hat{\sigma}^+ = |e\rangle\langle g|$ and $\hat{\sigma}^- = |g\rangle\langle e|$ are defined in terms of the ground and excited atomic states $|g\rangle$ and $|e\rangle$. The Hamiltonian $\hat{H}$ can be rewritten in the form

$$\hat{H} = \hat{H}_I + \hat{H}_{II}, \tag{A.11}$$

where

$$\hat{H}_I = \hbar\omega_c (\hat{a}^+ \hat{a} + \frac{\hat{\sigma}_z}{2}), \tag{A.12}$$

$$\hat{H}_{II} = \hbar \frac{\delta \hat{\sigma}_z}{2} + \hbar \frac{\Omega}{2} (\hat{a}\hat{\sigma}^+ + \hat{a}^+ \hat{\sigma}^-), \tag{A.13}$$

with $\delta = \omega_a - \omega_c$, are easily seen to commute. In turn, the eigenstates of $\hat{H}_I$ with a given number n of radiation quanta are given by $|n\rangle|g\rangle$, $|n\rangle|e\rangle$ and the states $|\psi_{1n}\rangle \equiv |n\rangle|e\rangle$, $|\psi_{2n}\rangle \equiv |n+1\rangle|g\rangle$ are degenerate with respect to $\hat{H}_I$. The matrix elements of the total Hamiltonian $\hat{H}$ in the subspaces $\{|\psi_{1n}\rangle, |\psi_{2n}\rangle\}$ read

$$H^{(n)} = \hbar \begin{pmatrix} n\omega_c + \frac{\omega_a}{2} & \frac{\Omega}{2}\sqrt{n+1} \\ \frac{\Omega}{2}\sqrt{n+1} & (n+1)\omega_c - \frac{\omega_a}{2} \end{pmatrix}. \tag{A.14}$$

Hereafter, we are interested in the case in which no more than one photon is present, i.e., we consider only $H^{(0)}$ whose eigenvalues turn out to be

$$E_{\pm} = \frac{\hbar\omega_c}{2} \pm \frac{\hbar\Omega_0(\delta)}{2}, \tag{A.15}$$

($\Omega_0(\delta) = \sqrt{\delta^2 + \Omega^2}$). The corresponding eigenstates are

$$|+\rangle = \cos(\alpha/2)|\psi_{1n}\rangle + \sin(\alpha/2)|\psi_{2n}\rangle, \tag{A.16}$$

$$|-\rangle = -\sin(\alpha/2)|\psi_{1n}\rangle + \cos(\alpha/2)|\psi_{2n}\rangle, \tag{A.17}$$

with $\alpha = \arctan(\Omega/\delta)$. We now consider the situation in which, at time $t = 0$, the atom is in the ground state in the presence of one photon. Therefore, since $|+\rangle$, $|-\rangle$ are stationary states with respective energies E_+ and E_-, we have

$$\begin{aligned} |\psi(t)\rangle = \\ \cos(\alpha/2)\,[\cos(\alpha/2)|1\rangle|g\rangle - \sin(\alpha/2)|0\rangle|e\rangle]\,e^{-iE_- t/\hbar} \\ + \sin(\alpha/2)\,[\cos(\alpha/2)|0\rangle|e\rangle + \sin(\alpha/2)|1\rangle|g\rangle]\,e^{-iE_+ t/\hbar}, \end{aligned} \tag{A.18}$$

where $|\psi(t)\rangle$ is the state of the system atom-plus-photon satisfying the required initial conditions.

We now consider a photon with angular frequency ω_c sent on a beam splitter, after which it can reach at time $t = 0$ either a cavity A or a cavity B, tuned to ω_c. In both cavities there is an atom with transition frequency ω_a. Initially, both atoms A and B are in the ground state, and the two cavities are far enough that no direct atomic interaction takes place. Thus, the initial state $|\psi_T(0)\rangle$ of the total system atom A + atom B + photon reads

$$|\psi_T(0)\rangle = \left(\frac{|1_A, 0_B\rangle + |0_A, 1_B\rangle}{\sqrt{2}}\right)|g_A, g_B\rangle, \tag{A.19}$$

with self-explanatory symbols. As a consequence, we readily obtain, with the help of Eq. (A.19),

$$\begin{aligned} |\psi_T(t)\rangle = \\ = [\frac{1}{\sqrt{2}}\cos^2(\alpha/2)|1_A, 0_B\rangle|g_A, g_B\rangle - \frac{1}{\sqrt{2}}\cos(\alpha/2)\sin(\alpha/2)|0_A, 0_B\rangle|e_A, g_B\rangle]e^{-iE_- t/\hbar} \\ + [\frac{1}{\sqrt{2}}\sin(\alpha/2)\cos(\alpha/2)|0_A, 0_B\rangle|e_A, g_B\rangle + \frac{1}{\sqrt{2}}\sin^2(\alpha/2)|1_A, 0_B\rangle|g_A, g_B\rangle]e^{-iE_+ t/\hbar} \\ + [\frac{1}{\sqrt{2}}\cos^2(\alpha/2)|1_B, 0_A\rangle|g_B, g_A\rangle - \frac{1}{\sqrt{2}}\cos(\alpha/2)\sin(\alpha/2)|0_B, 0_A\rangle|e_B, g_A\rangle]e^{-iE_- t/\hbar} \end{aligned}$$

$$
\begin{aligned}
&+[\frac{1}{\sqrt{2}}\sin(\alpha/2)\cos(\alpha/2)|0_B,0_A\rangle|e_B,g_A\rangle+\frac{1}{\sqrt{2}}\sin^2(\alpha/2)|1_B,0_A\rangle|g_B,g_A\rangle]e^{-iE_+t/\hbar}\\
&=[\cos^2(\alpha/2)e^{-iE_-t/\hbar}+\sin^2(\alpha/2)e^{-iE_+t/\hbar}]\left(\frac{|1_A,0_B\rangle+|0_A,1_B\rangle}{\sqrt{2}}\right)|g_A,g_B\rangle\\
&+\sin(\alpha/2)\cos(\alpha/2)\left(e^{-iE_+t/\hbar}-e^{-iE_-t/\hbar}\right)|0_A,0_B\rangle\left(\frac{|e_A,g_B\rangle+|g_A,g_B\rangle}{\sqrt{2}}\right)\\
&\equiv|\psi_D(t)\rangle+|\psi_E(t)\rangle,
\end{aligned}
\tag{A.20}
$$

where $|\psi_D(t)\rangle$ (both atoms in the ground state $\rightarrow|g_A,g_B\rangle$) forbids entanglement between the atoms A and B, while $|\psi_E(t)\rangle$ (photon disappeared $\rightarrow|0_A,0_B\rangle$) admits entanglement. On the other hand we have, with the help of Eq. (A.15),

$$
\langle\psi_D(t)|\psi_D(t)\rangle=\cos^4\left(\frac{\alpha}{2}\right)+\sin^4\left(\frac{\alpha}{2}\right)+2\sin^2\left(\frac{\alpha}{2}\right)\cos^2\left(\frac{\alpha}{2}\right)\cos(\Omega_0 t)\,, \tag{A.21}
$$

$$
\langle\psi_E(t)|\psi_E(t)\rangle=2\cos^2\left(\frac{\alpha}{2}\right)\sin^2\left(\frac{\alpha}{2}\right)[1-\cos(\Omega_0 t)]\,. \tag{A.22}
$$

Thus, total entanglement requires $\alpha=\arctan(\Omega/\delta)\simeq\pi/2$, for which

$$
\langle\psi_D(t)|\psi_D(t)\rangle=\frac{1}{2}[1+\cos(\Omega t)]\,, \tag{A.23}
$$

$$
\langle\psi_E(t)|\psi_E(t)\rangle=\frac{1}{2}[1-\cos(\Omega t)]\,. \tag{A.24}
$$

The condition $\langle\psi_E(t)|\psi_E(t)\rangle=1$ and $\langle\psi_D(t)|\psi_D(t)\rangle=0$ emerges for times t_n given by

$$
t_n=\frac{(2n+1)\pi}{\Omega_0}. \tag{A.25}
$$

Moreover, the condition $\alpha\simeq\pi/2$ implies that $\delta/\Omega_0\ll 1$. However, δ cannot be less than the intrinsic uncertainty $\Delta\omega$ of the photon frequency, so that the photon must have an uncertainty in time $\Delta t>2\pi/\Delta\omega\geq 2\pi/\delta$. On the other hand, to use Eq. (A.22) we need a resolution in time Δt such that $\Delta t<2\pi/\Omega_0$, so that, since $\Omega_0=(\delta^2+\Omega^2)^{1/2}$, $\Delta t\ll 2\pi/\delta$, which does not agree with the minimum uncertainty Δt, i.e., $\Delta t\geq 2\pi/\delta$. Specifically, the condition on the required time resolution contradicts the uncertainty in the arrival times of the photon on the two atoms. On the physically available time resolution Δt experiments yield the time averages, i.e., in Eqs. (A.23) and (A.24), the time t possesses an indetermination Δt. Since this limit is a consequence of the uncertainty principle, it implies that our formulation of Eq. (A.20) as parametrized in time when $\alpha\simeq\pi/2$ is incorrect, and the true wavefunction that describes our system must not contain t, i.e., be

$$
|\psi_T\rangle=\frac{1}{\sqrt{2}}[|\psi_D\rangle+|\psi_E\rangle]\,, \tag{A.26}
$$

where

$$|\psi_D\rangle = \frac{1}{\sqrt{2}}\left[|1_A, 0_B\rangle + |0_A, 1_B\rangle\right]|g_A, g_B\rangle, \tag{A.27}$$

$$|\psi_E\rangle = \frac{1}{\sqrt{2}}|0_A, 0_B\rangle\left[|e_A, g_B\rangle + |g_A, e_B\rangle\right]. \tag{A.28}$$

Formally, our attempt to entangle one particle using two distant systems is achieved through an appropriate interaction such that the prepared input state $|1\rangle_p|\alpha\rangle_A|\beta\rangle_B$ evolves to the final state $|0\rangle_p|\gamma\rangle_{AB}$, with the constraint that $|\gamma\rangle_{AB}$ is entangled. The limits associated to the use of one particle to entangle two distant systems can, in this vein, also be analyzed considering the details of the interaction Hamiltonian that represents the evolution $|0\rangle_p|\gamma\rangle_{AB} = \hat{a}\hat{P}|1\rangle_p|\alpha\rangle_A|\beta\rangle_B$, where $\hat{a}$ is the annihilation operator for p, and $\hat{P}$ is the operator that describes the evolution for the systems. Assume now that no interaction term involves the entangling particle and the two systems together. This means that, using for example the standard spinorial notation for two two-level systems A and B, $\hat{P} = \hat{P}(\hat{\sigma}_{z,A}, \hat{\sigma}_{z,B})$ is a linear combination of $\hat{\sigma}_{z,A}, \hat{\sigma}_{z,B}$. Now, the maximally entangled Bell-basis are the eigenstates of nonlinear operators in the spin of the two systems, such as $\hat{\sigma}_z^2 = (\hat{\sigma}_{z,A} + \hat{\sigma}_{z,B})^2$ and $\hat{\sigma}_x^2 = (\hat{\sigma}_{x,A} + \hat{\sigma}_{x,B})^2$, and the latter does not commute with $\hat{\sigma}_{z,A}, \hat{\sigma}_{z,B}$, and hence with $\hat{P}$ [2]. Specific combinations of $\hat{\sigma}_{x,A}$ and $\hat{\sigma}_{x,B}$ can share eigenstates with the disentangled basis, and a maximum overlap of half the Hilbert space occurs when the interaction involves the total spin operators of A and B. This is the case of the previously discussed Jaynes–Cummings interaction, where the Hamiltonian has terms with $\hat{\sigma}_i = \hat{\sigma}_{i,A} + \hat{\sigma}_{i,B}$, $(i = x, y, z)$. The requirement that the particle system interaction involves nonlinear operators, such as $\hat{\sigma}^-\hat{\sigma}^+ + \hat{\sigma}^+\hat{\sigma}^-$, where $\hat{\sigma}^\pm = \hat{\sigma}_A^\pm + \hat{\sigma}_B^\pm$, and $\hat{\sigma}_{A/B}^\pm = \hat{\sigma}_{x,A/B} \pm i\hat{\sigma}_{y,A/B}$, implies that a three-particle interaction term must be involved, that is, that the two systems must locally interact and separate subsequently (i.e., they cannot be prepared as "distant"), as is well known to occur for standard entanglement sources.

A.1.2 Remarks on the Conditional Nature

These results indicate that one can achieve conditional maximum 50% entanglement between two distant systems using a single particle.[1] Conditional means that of all the results on $(\hat{\sigma}_{i,A}, \hat{\sigma}_{i,B})$ non-local correlations can only be detected in post-selected ensembles. This may be misinterpreted as being a practical limitation, and hence that total entanglement can actually be distilled from this conditional ensemble before measurements are carried out. In order to sift an entangled state from $|\psi_T\rangle$, we can enclose the outer surface of each of the two cavities A and B with two ideal photon detectors D_A, D_B (Fig. A.2), and simultaneously open A and B at a given instant of

[1]The same factor of 50% has been obtained in the field of linear teleportation [1, 4, 6].

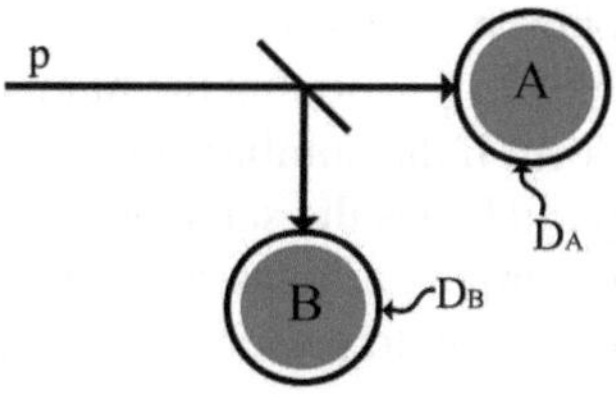

Fig. A.2 Entangling system

time ($t = 0$). If either D_A or D_B clicks (at $t = 0$), $|\psi_T\rangle$ is projected (at $t = 0$) onto $|\psi_D\rangle$, whose atomic component

$$|\phi_D(t = 0)\rangle = |g_A, g_B\rangle \tag{A.29}$$

is disentangled, while, if neither D_A nor D_B clicks, the mixed state is projected onto the state $|\psi_E\rangle$, whose atomic component

$$|\phi_E(t = 0)\rangle = \frac{1}{\sqrt{2}} [|e_A, g_B\rangle + |g_A, e_B\rangle] \tag{A.30}$$

is entangled. We note that this result is obviously valid whenever the typical time T_C necessary to cross the cavity is much smaller than the oscillation time $2\pi/\Omega$. This is the case for the example referring to hydrogen atoms in a cavity of about $V = 10^{-9}m^3$, for which, with the help of Eq. (A.10), $2\pi/\Omega = 4 \cdot 10^{-6}s$, while $T_C \simeq \frac{V^{1/3}}{c} \simeq \frac{10^{-11}}{3}s$. In fact, the validity of Eq. (A.30) hides a more subtle but profound assumption. We have an entangled pair of systems only once we have certainty that both detectors D_A and D_B have not fired. This shared knowledge at A and B can be readily achieved assuming that the time it takes a classical signal to reach from A to B would be smaller than the characteristic time $2\pi/\Omega$ in which the system oscillates. This apparently trivial condition means that for any given systems A and B, there is a distance beyond which the entangling scheme breaks down. The existence of this distance means that the two systems A and B obey Einstein locality [3].

A.1.3 Remarks on Reversibility

The entire matter as to if and how a specific process can or cannot lead to entanglement can be analyzed in terms of microscopic reversibility. Specifically, consider a process in which an isolated system with N independent input particles leads to N output particles at space-like distances. Should the final wavefunction be entangled, the outcomes of the N independent experiments that can be carried out on the N output particles would be correlated, so that the phase-space of the outcoming state would be smaller than that of the incoming state, thus violating microscopic reversibility. Congruently, the Jaynes–Cummings model associated with the split photon interacting with the two single atoms described by Eq. (A.4) contains terms like $\hat{a}\hat{\sigma}^{+}_{A/B}$, so that the number of particles is constant. Thus, entanglement can only occur if the number of particles from input to output changes. More precisely, if each particle has the same number of internal states both at input and output, the number of output particles susceptible to space-like experiments must increase, as this is the only way to preserve the phase-space of the system and introduce correlations.

References

1. Bouwmeester D, Pan J-W, Mattle K, Eibl M, Weinfurter H, Zeilinger A (1997) Experimental quantum teleportation. Nature 390(6660):575–579
2. DelRe E, Crosignani B, Di Porto P (2000) Scheme for total quantum teleportation. Phys Rev Lett 84(13):2989
3. Einstein A, Podolsky B, Rosen N (1935) Can quantum-mechanical description of physical reality be considered complete? Phys Rev 47(10):777
4. Lütkenhaus N, Calsamiglia J, Suominen K-A (1999) Bell measurements for teleportation. Phys Rev A 59(5):3295
5. Shore BW, Knight PL (1993) The jaynes-cummings model. J Mod Opt 40(7):1195–1238
6. Vaidman L, Yoran N (1999) Methods for reliable teleportation. Phys Rev A 59(1):116
7. Van Enk SJ (2005) Single-particle entanglement. Phys Rev A 72(6):064306

Curriculum Vitae

Giuseppe Di Domenico Ph.D
May 16th, 1988
Via Tiburtina 68, 00185 Rome, Italy.
+39 392-9260097 (ITA)
+972 58-627-8903 (IL)
www.linkedin.com/in/giuseppe-didomenico
giuseppedidomenico8@gmail.com

Research Interests

My main interests focuses on non-linear optics; particular topics where I've done research include: frequency mixing and refractive index engineering, spatial optical solitons, electroholography, para to ferroelectric transition and imaging in turbid media. I'm also interested quantum optics and in connections between nonlinear optics and schemes for quantum entanglement.

G. Di Domenico, *Electro-optic Photonic Circuits*, Springer Theses,
https://doi.org/10.1007/978-3-030-23189-7

Working Experience

since 2018 Research Associate Post-Doc
Currently I am employed by the department of electrical engineering in Tel Aviv University TAU (Israel). My main research topic is the spatial and spectral shaping of spontaneously down-converted photons, I am also interested in device for mode demultiplexing of optical signals.

2013–2014 Traineeship Apprentice
Eindhoven University of Technology (TUE). Eindhoven, Nederland. Internship and lab training work related to my master thesis. My main activity was the experimental study of the char combustion in a fluidized bed reactor.

Education

2014–2018 Ph.D in Physics Ph.D
Ph.D Candidate at "La Sapienza" University of Rome, Department of Physics. Thesis title: "Electro-optic photonic circuits from linear and nonlinear waves in nanodisordered photorefractive ferroelectrics".

2011–2014 Chemical Engineering Processes Master Degree
Master degree in Chemical Engineering obtained from Universitá degli studi di L'Aquila in L'Aquila (AQ), Thesis title: "Simulation of the behaviour of a perovskite membrane for char combustion in fluidized bed", mark 110/110.

2007–2011 Chemical Engineering Bachelor Degree
Bachelor graduation obtained from Universitá degli studi di L'Aquila in L'Aquila (AQ), Thesis title: "Quantum Techniques applied to Sorting Algorithms", mark 101/110.

2002–2007 Secondary School focusing on Sciences High School
Baccalaureate (Compared to English A level) obtained from Liceo Scientifico G.Galilei in Riccia (CB), mark 92 /100.

Publications

2018 **Sci. Rep. 8**, 1 DOI Link
G. Di Domenico, G. Ruocco, C. Colosi, E. DelRe and G. Antonacci. "Self-suppression of Bessel beam side lobes for high-contrast light sheet microscopy."

2018 **Phys. Rev. A. 98**, 3 DOI Link
M. Flammini, G. Di Domenico, D. Pierangeli, F. Di Mei, A.J. Agranat and E. DelRe.
"Observation of Bessel-beam self-trapping."

2017 *Sci. Rep.* **7**, 17 DOI Link
G. Antonacci[a], G. Di Domenico[a], S. Silvestri, E. DelRe and G. Ruocco.
"Diffraction-free light droplets for axially-resolved volume imaging."

2017 *Appl. Opt.* **56**, 2908-2911 DOI Link
G. Di Domenico, J. Parravicini, G. Antonacci, S. Silvestri, A.J. Agranat and E. DelRe.
"Miniaturized photogenerated electro-optic axicon lens Gaussian-to-Bessel beam conversion."

2017 *Opt. Lett.* **42**, 3856 DOI Link
M. Ferraro, D. Pierangeli, M. Flammini, G. Di Domenico, L. Falsi, F. Di Mei, A. J. Agranat, and E. DelRe,
"Observation of polarization-maintaining light propagation in depoled compositionally disordered ferroelectrics."

2016 *Nat. Comm.* **7**, 10674 DOI Link
D. Pierangeli, M. Ferraro, F. Di Mei, G. Di Domenico, C.EM. De Oliveira, A.J. Agranat and E. DelRe. "Super-crystals in composite ferroelectrics."

2016 *Phys. Rev Lett.* **116**, 153902 DOI Link
F. Di Mei, P. Caramazza, D. Pierangeli, G. Di Domenico, H. Ilan, A.J. Agranat, P. Di Porto and E. DelRe. "Intrinsic negative mass from nonlinearity."

2016 *Phys. Rev Lett.* **117**, 183902 DOI Link
D. Pierangeli, F. Di Mei, G. Di Domenico, A.J. Agranat, C. Conti, and E. DelRe. "Turbulent transitions in optical wave propagation."

2016 *Phys Rev. App. 6*, 054020 DOI Link
G. Antonacci, S. De Panfilis, G. Di Domenico, E. DelRe and G. Ruocco.
"Breaking the Contrast Limit in Single-Pass Fabry-Pérot Spectrometers."

2016 **Phys. Rev. A 94**, 063833 DOI Link
D. Pierangeli, G. Musarra, F. Di Mei, G. Di Domenico, A.J. Agranat, C. Conti and E. DelRe. "Enhancing optical extreme events through input wave disorder."

(a: Equal contribution)

Conference Paper

2018 CLEO Conference on Lasers and Electro-Optics (OSA). DOI Link
M. Flammini, G. Di Domenico, D. Pierangeli, F. Di Mei, A. J. Agranat, and E. DelRe.
"Observation of a Bessel beam soliton."

2017 Focus On Microscopy. DOI Link
G. Antonacci, D. Di Domenico, S. Silvestri, E. Delre and G. Ruocco.
"Diraction-Free and Self-Healing Light Droplets for Deep Volume Imaging."

2017 CLEO Conference on Lasers and Electro-Optics (OSA). DOI Link
G. Antonacci, S. De Panfilis, G. Di Domenico, E. DelRe, and G. Ruocco.
"A 1000-fold contrast enhancement in Fabry-Pérot interferometers."

2016 CLEO Conference on Lasers and Electro-Optics (OSA). DOI Link
D. Pierangeli, M. Ferraro,F. Di Mei, G. Di Domenico, C.E.M De Oliveira, A.J. Agranat and E. DelRe.
"Spontaneous photonic super-crystal in composite ferroelectrics." *(oral presentation)*

Other Information

Languages Italian (Mother tongue)
English C2 (Certificate B1)

Programming Python, Matlab, Matemathica, LaTeX, basic knowledge of other programming languages.

Driving licence B

MIX
Papier aus verantwortungsvollen Quellen
Paper from responsible sources
FSC® C105338

If you have any concerns about our products,
you can contact us on
ProductSafety@springernature.com

In case Publisher is established outside the EU,
the EU authorized representative is:
Springer Nature Customer Service Center GmbH
Europaplatz 3, 69115 Heidelberg, Germany

Printed by Libri Plureos GmbH
in Hamburg, Germany